Magnus Troilius

NOTES on the CHEMISTRY of IRON

Magnus Troilius

NOTES on the CHEMISTRY of IRON

ISBN/EAN: 9783741122125

Manufactured in Europe, USA, Canada, Australia, Japa

Cover: Foto ©Angelika Wolter / pixelio.de

Manufactured and distributed by brebook publishing software (www.brebook.com)

Magnus Troilius

NOTES on the CHEMISTRY of IRON

NOTES

ON THE

CHEMISTRY OF IRON.

FOR PROFESSIONAL MEN, STUDENTS, IRON AND STEEL MERCHANTS,
AND ALL INTERESTED IN IRON.

BY

MAGNUS TROILIUS, E.M.

SECOND EDITION, REVISED AND ENLARGED.

NEW YORK:
JOHN WILEY & SONS, PUBLISHERS.
1886.

TO THE MEMORY

OF

THE LATE MR. ALEX. L. HOLLEY

THIS BOOK

IS

DEDICATED.

" He visits America . . . to see what is doing . . . and to contribute, if he can, in the matter of a uniform system of analysis of steel."

(Extract from a general letter of introduction written for the Author by Mr. Holley in London, December 12, 1881.)

PREFACE.

IN presenting this little work to the public, the author has endeavored to embody in plain language a thoroughly practical description of such chemical methods of analysis in iron and steel manufacture as have come under his personal observation during some years of practice.

It is hoped that their special adaptation to their purpose will be at once recognized, and their combination of rapidity with accuracy noted.

Wrought iron and steel, being of special interest at this time, have been particularly prominent in the work.

The methods of analysis having been explained, the reader is also shown how to apply the results obtained; and thus the practical character of the work is at once established.

The thanks of the author are due to Mr. R. W. Davenport, Superintendent of the Midvale Steel Company, for valuable advice; and to Mr. E. Hallgren, C.E., for labor bestowed on the drawings.

It is hoped that the principles and practice here set forth will prove of assistance to iron and steel manufacturers and better enable them to meet the severe demands made upon their products in this age of progress. If such be in any measure the result of his efforts, none will be better pleased than the

AUTHOR.

PREFACE TO THE SECOND EDITION.

THE first edition being exhausted, a second was called for and is herewith published.

Several additions and enlargements have been made, and the Author wishes to express his thanks to all those who have lent their friendly criticism and advice in working out the same.

Mr. O. L. Kehrwieder, late my assistant at the Midvale Steel Company, has contributed the list of requisites for an Iron Laboratory, and also successfully worked out the details of the method for determining aluminium in alloys with iron, etc., suggested by the Author.

A list is given of chemists whose labors have been more or less associated with some of the methods described in this book.

The chapter on Gas Analysis has been extended; chapters on Electrolysis and Metallurgical Applications have been added, and the Appendices have received numerous additions.

MAGNUS TROILIUS.

PHILADELPHIA, *March*, 1886.

CONTENTS.

CHAPTER I.

GENERAL REMARKS REGARDING THE DISTINCTIVE PROPERTIES OF PIG-IRON, WROUGHT IRON AND STEEL, AND THE INFLUENCE OF THE VARIOUS ELEMENTS USUALLY COMBINED AND ALLOYED WITH THE SAME 1

CHAPTER II.

DETERMINATION OF ELEMENTS MOST FREQUENTLY OCCURRING COMBINED OR ALLOYED WITH IRON.

A. Analysis of Wrought Iron and Steels 7
   a. Determination of Carbon by Combustion in Oxygen . 7
   b. Determination of Carbon by Combustion with Chromic and Sulphuric Acids 16
   c. Other Combustion Methods for Carbon . . . 18
   d. The Determination of Carbon by Color . . . 19
   e. Determination of Phosphorus 23
   f. Determination of Manganese 29
   g. Determination of Silicon 35
   h. Determination of Sulphur 36
   i. Determination of Copper 39
   j. Determination of Slag and Oxide of Iron . . . 41
   k. Determination of Arsenic 44
   l. Determination of Titanium 45
   m. Tracing of Vanadium 46
   n. Determination of Chromium 46
   o. Determination of Tungsten 48

B. Analysis of Pig-iron 49
a. Determination of Graphite 49
b. Determination of Silicon 49
c. Determination of Sulphur 50
d. Slag 50
C. Analysis of Spiegel and Ferromanganese 50
D. Analysis of Silicon-iron, etc. 53

CHAPTER III.

Determination of the Most Important Ingredients in Iron Ores, Slags, Limestones, Fuel, etc.

A. Analysis of Iron Ores 55
    a. Determination of the Total Iron 55
    b. Determination of the Iron present as Ferric Oxide . 57
    c. Determination of Phosphorus 58
    d. Determination of Sulphur 59
    e. Determination of Manganese 60
    f. Determination of Moisture and Loss on Ignition . 60
    g. Complete Analysis of Iron Ores 61
    h. Elements of Rare Occurrence 68
    i. The Dry Assay of Iron Ores 70
B. Analysis of Slags, etc. 72
C. Analysis of Coal and Coke 73

CHAPTER IV.

Notes on Gas Analysis 76

CHAPTER V.

Metallurgical Notes and Practical Uses of the Results of Analyses 95

CHAPTER VI.

Notes on Electrolysis 107

CONTENTS. ix

	PAGE
APPENDICES	113
A. Heat Calculations	115
B. Calculation of Blast-furnace Burden	121
C. Table for Rapid Calculation of Analyses	124
D. Etching Test	125
E. Table of Elements	126
F. French Weights and Measures	127
G. "Body in Steel"	129
H. Melting Points, etc.	131
I. Laboratory Requisities	132
J. Iron Ores	137
K. Organic Matter in Water	139
L. Table Giving the Tension of Aqueous Vapor	141
M. Useful Tables	142

NOTES ON THE CHEMISTRY OF IRON.

CHAPTER I.

GENERAL REMARKS REGARDING THE DISTINCTIVE PROPERTIES OF PIG-IRON, WROUGHT IRON, AND STEEL, AND THE INFLUENCE OF THE VARIOUS ELEMENTS USUALLY COMBINED AND ALLOYED WITH THE SAME.

By pig-iron we mean the metal obtained by the reduction of iron ores in the blast-furnace. Pig-iron is used as raw material in the manufacture of wrought iron and steel: it is also largely used for making castings. The chemical composition of pig-iron varies according to the purposes for which it is intended. In the blast-furnace are also manufactured various alloys of iron with manganese (spiegel, ferromanganese), and alloys of iron with silicon, or with both silicon and manganese ("ferro silicium," "special pig").

By wrought iron we mean the metal obtained from processes such as the Swedish Lancashire, Catalan, and puddling, in which processes the refined metal does not occur in a fluid condition. Wrought iron always contains a considerable amount of slag and oxides of iron in mechanical admixture. By steel we mean the metal obtained from the crucible, Bessemer, Siemens, and other processes, where the refined metal is obtained in a fluid condition, and can be cast into moulds. Steel seldom contains more than a very small amount of oxide of iron and slag.

We have further to note so-called *malleable* iron castings. Such castings are made of pig-iron of suitable chemical composition. By heating such castings in oxidizing substances (oxides of iron, etc.), excess of carbon is eliminated, and the metal is rendered malleable.

The elements which most usually occur in pig-iron, wrought iron, steel, etc., are carbon, phosphorus, manganese, silicon, sulphur, copper, chromium, and tungsten. More seldom there is found arsenic, antimony, nickel, and cobalt. Chromium and tungsten mostly occur in the special steels named after said elements. Titanium and vanadium are occasionally found. Slag and oxide of iron are of importance in many cases. Iron and steel also contain more or less hydrogen, nitrogen, carbonic oxide, and other gases. For these we have no really practical methods of analysis.

As to the influence of the various elements above mentioned on the properties of iron and steel, it is only with regard to the eight first that we possess much accurate knowledge. No unvarying rules, however, can be laid down. The reason for this lies in the nature and conditions of the various processes of manufacture. Besides the many external and physical forces that tend to modify the influence of chemical composition, such as rolling, hammering, hardening, and annealing, there may be formed, at various temperatures and under various circumstances, innumerable chemical compounds of different molecular structure. These compounds may again split up into others, when changes occur in the said conditions. These phenomena are analogous to those so well known in the dissociation of gases. Castings often have different chemical composition in different parts. This is due to the separation of such compounds. When, therefore, we determine the total amounts of carbon, silicon, etc., in iron and steel, and try to define their relations to physical properties, we must bear in mind that we may not know the composition of the various compounds or alloys which said elements have formed in the metal, and that these compounds and alloys may be different in steels of the same chemical composition. Thus the physical properties of apparently similar steels may differ widely.

Iron forms alloys in *any* proportions with manganese, chromium, tungsten, nickel, cobalt, copper, gold, platinum, aluminium, antimony, silicon, sulphur, phosphorus, and arsenic: in *limited* proportions only with zinc, tin, bismuth, and carbon. It scarcely alloys at all with lead, silver, and mercury.

Influence of Carbon. We have to consider carbon as occurring in iron in at least two distinct forms or conditions, — " graphitic " and " combined carbon." Graphitic carbon occurs almost exclusively in gray pig-iron, taking the form of dark, thin flakes, varying much in size, and intersecting the molecules or small particles of iron. Its influence is to make the pig-iron softer and tougher than would be the case if it existed in the form of "combined carbon."

"Combined carbon" is now generally supposed to occur under two distinct forms. The one, so-called "cement carbon," consists of alloys of iron, carbon, and a little silicon, uniformly distributed or dissolved in the metal. The other is the so-called "hardening carbon." The cement carbon remains as a black residue when the metal is dissolved in cold acids, whilst the hardening carbon passes off as hydro-carbon gases of the ethylene ($C_2 H_4$) series. By heating iron or steel to a red heat, and rapidly cooling in water, the cement carbon is converted into hardening carbon. Simply hammering cold has the same effect to some extent. *Slow* cooling or annealing produces cement carbon. Doubtless these phenomena can be classed with those referred to above, as analogous to the dissociation of gases. It is therefore evident, that simply the knowledge of *the amount of total carbon* is not enough to explain the physical influence of the same. In general, we may say that the *tensile strength* of steel increases with the combined carbon up to about one per cent, again decreasing as the carbon increases above this point. The *elastic limit* is raised by combined carbon somewhat more rapidly than is the tensile strength.

Influence of Silicon. In the manufacture of pig-iron in the blast-furnace, silicon replaces carbon to some extent, and promotes the production of gray pig-iron, i.e., the formation of graphitic carbon. Otherwise its influence in iron and steel is similar to that of carbon, but less active.

Influence of Phosphorus. Phosphorus is considered to exist in irons and steels as the phosphide of iron ($Fe_4 P_2$). There are many other iron phosphides known; but the one represented by the above formula is the poorest in phosphorus of them all, and may, therefore, be considered as being dissolved in an excess of metallic iron.

Phosphorus causes cold-shortness, i.e., brittleness when cold; and the more so the greater the amount of carbon present. The presence of silicon, on the other hand, carbon being absent, appears to modify to a considerable degree the effect of phosphorus to cause coldshortness. Advantage is sometimes taken of this fact, when a large amount of phosphorus is present, to obtain hardness in steel by replacing carbon with silicon. Both carbon and silicon render the metal more fusible, but phosphorus does so in a much higher degree. Phosphorus, especially in steel, has a great tendency to render the metal crystalline.

Influence of Sulphur. Sulphur occurs in iron and steel as the monosulphide of iron ($Fe\ S$). Sulphur opposes the absorption of carbon in the blast-furnace process, but promotes the formation of combined carbon. Sulphur makes wrought iron and steel red-short, i.e., brittle at a red heat; but its direct influence on the cold metal when present in moderate quantities is doubtful.

Influence of Manganese. Manganese aids the absorption of carbon in the blast-furnace and the formation of combined carbon. The influence of manganese on wrought iron and steel varies with the other ingredients. It seems to act like carbon in most cases, though with less energy, and to make the metal more easily worked at a forging heat. It is thus a remedy for sulphur, whilst it acts indifferently, like silicon, towards phosphorus. In the manufacture of steel, manganese is added at the end of the process to remove oxygen from the bath. An addition of both manganese and silicon prevents blow-holes in steel.

Influence of Copper. Not much is known regarding the effect of this element. Some red-shortness seems to be caused by the presence of a high percentage of copper.

Oxygen in Steel causes red-shortness, and is considered to exist chiefly as the iron mon-oxide (Fe O). In wrought iron it occurs in the slags of various composition that enter into that metal.

Influence of Chromium and Tungsten. Both these elements render the metal very hard. Chromium has more effect than tungsten. Chromium steel seldom contains more than one per cent of chromium, whilst tungsten steel often has nine per cent of tungsten. Both chromium and tungsten steels are mostly used for cutting tools, etc. Chromium steel is also occasionally used for structural purposes.

CHAPTER II.

DETERMINATION OF ELEMENTS MOST FREQUENTLY OCCURRING COMBINED OR ALLOYED WITH IRON.

A.—Analysis of Wrought Iron and Steels.

THE analysis of wrought iron and steels is first considered, because the greatest possible accuracy is therein required and the methods must, therefore, be described in full detail. These same methods apply as a rule to the analysis of pig-irons, spiegel, ferromanganese, etc. The slight modifications that are necessary will be subsequently mentioned.

a. **Determination of Carbon by Combustion in Oxygen.** Five grams of drillings are weighed out and put into a flask of somewhat more than three hundred cubic centimetres capacity. The flask should be of thick glass. It must have a ground stopper and lip to facilitate pouring. A few drops of ammonia are put into the flask before putting the drillings into the same. The drillings must be prepared with care so as to contain no combustible matter. They should be extracted with a magnet when weighing out. Upon the drillings is poured a solution of the double chloride of copper and ammonia $\left(\begin{matrix} Cu \\ (NH_4)_2 \end{matrix} \right\} Cl_4 \right)$. This solution is prepared by dissolving one part by weight of the salt in three parts of water, filtering it through a purified asbestos filter and adding ammonia to the filtrate until a slight permanent precipitate is obtained. Three hundred cubic centimetres of this solution, with the green permanent precipitate well shaken up in it, is used for five grams of drillings. The stopper is inserted and the flask kept in agitation until all the separated copper has completely disappeared. The solution then becomes very neutral and a basic salt separates which is readily dissolved in a few drops of hydrochloric acid. The residue contains all the carbon, some silica, and possibly some iron-phosphide, etc. When a large number of determinations have to be made it is convenient to have many flasks marked differently. The flasks may then be placed in a shaking apparatus (Fig. I.), which consists of a board with holes, into which the flasks are wedged. This board is supported by uprights, one at each corner. The uprights have universal joints at both ends, so that the board can be agitated in a circular direction at a rapid rate without revolving or twisting in any way. Under the board and attached to it at the centre is a projecting pin moving in an arm. This arm is attached to a vertical spindle which, when revolved by means of a crank and bevelled gears, agitates the board. The arm has a perforation by means of which the projecting pin under the board is held at about three inches from the centre of the spindle. The apparatus may be driven by a small water or gas motor or by hand. By using

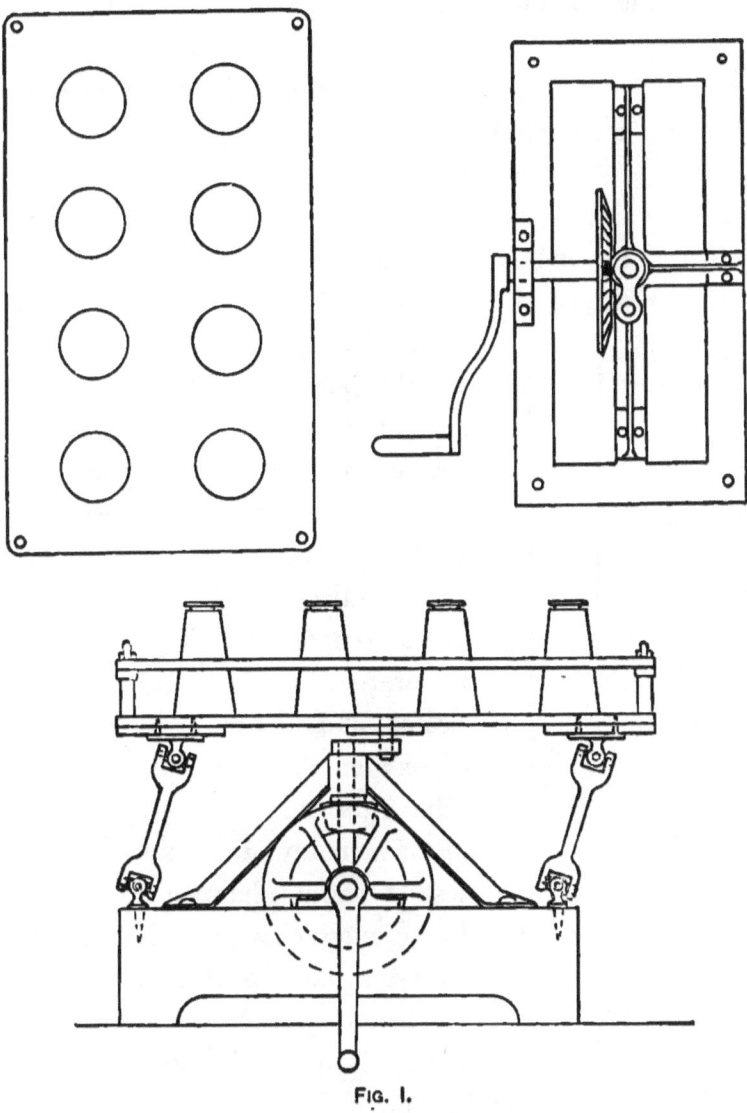

FIG. I.

this apparatus a large number of samples may be completely dissolved in less than one hour, particularly if the borings be fine. The reaction is principally —

$$Fe + 2\,Cu\,Cl_2 = Fe\,Cl_2 + Cu_2\,Cl_2.$$

Some copper separates at first but is afterwards redissolved on shaking:

$$Fe + Cu\,Cl_2 = Fe\,Cl_2 + Cu.$$
$$Cu + Cu\,Cl_2 = 2\,Cu\,Cl.$$

After allowing at least one hour for the settling of the carbon residues, they are filtered off into platinum funnels, of shape shown by Fig. II. At the bottom of the funnel is a coil of platinum wire to support the asbestos filter. This asbestos must be purified before using by washing with hydrochloric acid and igniting, — best in a current of oxygen or air. A good way is to prepare a filter of impure asbestos and ignite it in a stream of oxygen in the combustion-tube. The filter once thus prepared will last for many determinations, unless it should happen to become badly clogged. In this case the asbestos may be shaken out, loosened, and put back again. The funnel is inserted into one of the necks of a two-necked flask by means of a rubber-stopper and suction applied through the other neck

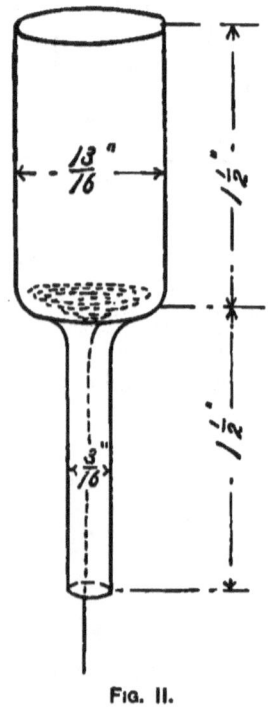

Fig. II.

by means of a good water suction-pump. By means of a glass rod the filter is packed carefully and tightly and filtering proceeded with. When the residue reaches

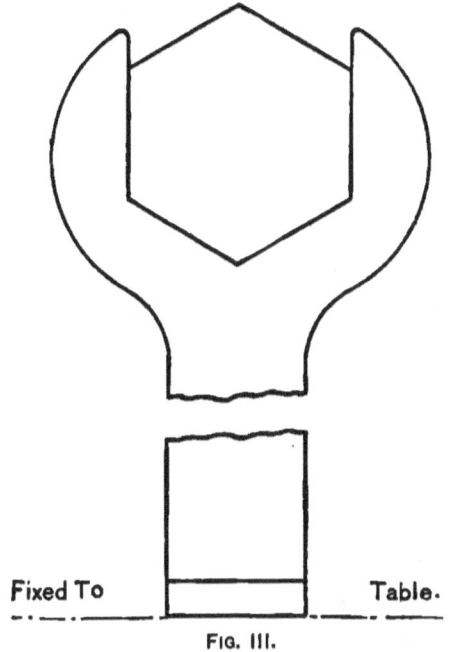

FIG. III.

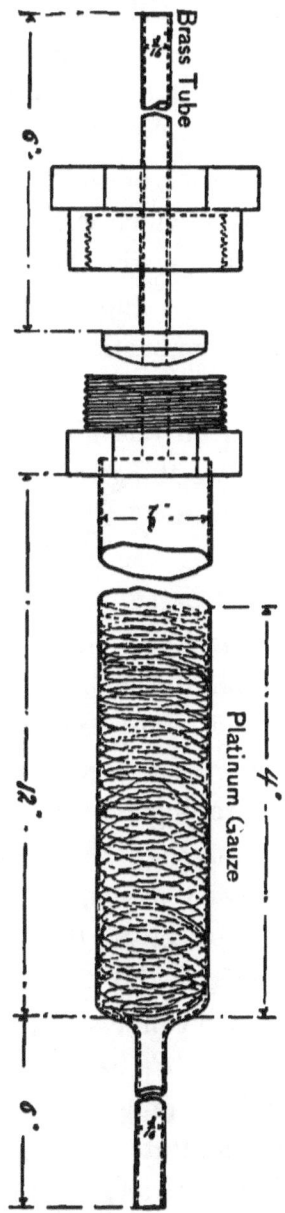

the filter the operation is often much retarded owing to the clogging properties of some carbon residues. With good suction and some practice, however, this does not cause serious difficulty. The residue is first washed with a little hydrochloric acid and finally with plenty of hot water. The acid must be used before using water. Water alone causes a white precipitate of sub-chloride of copper which clogs the filter badly. The platinum funnels with their contents are dried at 100° C. in an ordinary drying-box. When dry, — which is indicated by the shrinking of the black mass on the filter, — the residues are ready for combustion. Fig. IV. shows the general arrangement of the combustion-apparatus, and Fig. III. shows the correct dimensions of the combustion-tube (of platinum), with an air-tight joint of brass at one end.

Two things are essential for the successful use of this apparatus. Firstly, absolute purification of the oxygen and air. Secondly, absolute tightness of the apparatus in all parts. The first condition is fulfilled by passing oxygen and air through very capacious jars filled with sulphuric acid, fluid potash, and solid potash, respectively. The second condition can only be met by using the greatest care in fixing the rubber-stoppers and rubber-tubings on to the apparatus.

On Fig. IV. the oxygen receiver A, containing compressed oxygen, is seen standing on the floor. From this receiver the oxygen passes through the safety-bottle B into the bottle B_1 containing a solution of potash (three weights of potash in five weights of water). B and B_1 have a capacity of two liters each. Similarly arranged are the bottles C and C_1 containing concentrated sulphuric acid. They are each of one liter capacity. The safety-bottles prevent the liquids from mixing in case of back-pressure and from going down into the receiver. The oxygen finally passes up

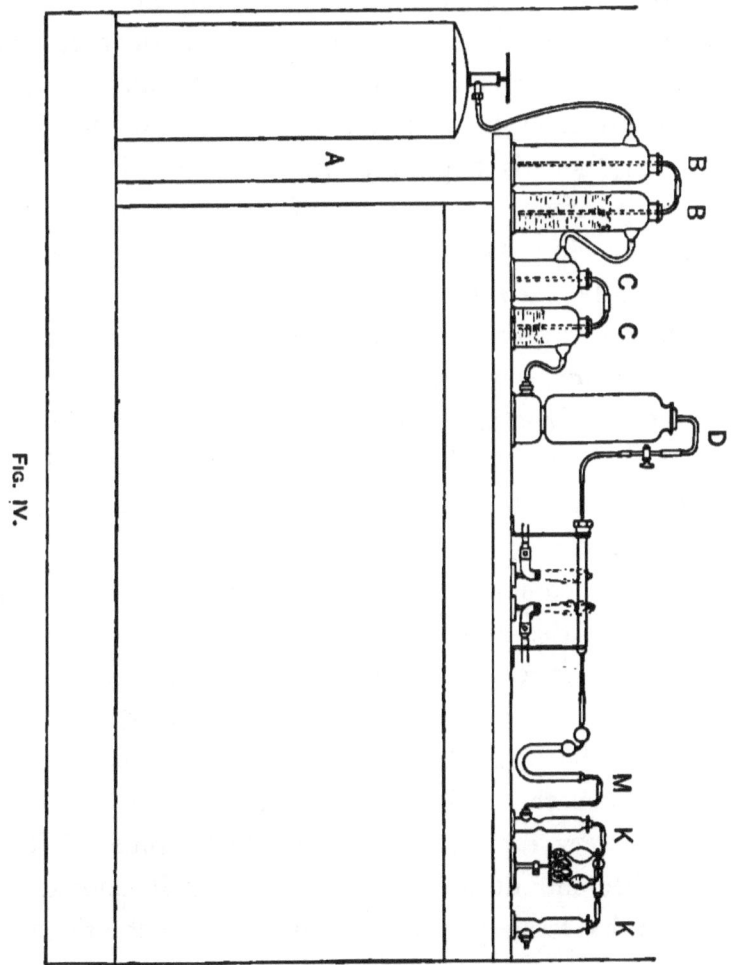

Fig. IV.

through a jar *D* of about two liters capacity, filled with pieces of solid potassium hydrate. This arrangement effects a very durable and reliable purification of the oxygen, freeing it from the last traces of moisture and carbonic acid. An exactly similar arrangement (not shown on Fig. IV.) is had for the air which is pressed through the apparatus after each combustion to expel the oxygen. A three-way glass tube connects the combustion-tube with the two potash-jars, for each of which there is a glass stop-cock; thus the oxygen and air purifiers can be shut off and connected alternately at will. Where blast is available it can be conveniently used to press air through the apparatus. It is better to press air through than to draw it through, the risk of drawing in impure air through some small leak being obviated. The platinum tube contains, at the end where its diameter is diminished, a roll of platinum gauze four inches long. This gauze insures the conversion of any carbonic oxide to carbonic acid during the combustion. The platinum tube is connected with the Marchand's tube M filled with small pieces of pumice-stone which have been thoroughly soaked with cupric sulphate and then heated to about 230° C. The white anhydrous cupric sulphate thus obtained absorbs moisture and traces of hydrochloric acid gas very rapidly, turning gradually blue. When it has become decidedly blue in color it should be replaced by fresh pumice. A little hydrochloric acid gas as well as some chlorides are generally volatilized during the combustion, even after the most thorough washing of the carbon residue. A little jar K filled with pieces of chloride of calcium is connected with the Marchand's tube. The *porous* calcium chloride should be used, not the *fused*, the latter being apt to contain some calcium oxide which absorbs carbonic acid. The *porous* calcium chloride offers more surface to the passing gases and absorbs moisture more quickly. Between this jar and the similar jar K_1, — the so-called safety-jar filled with pieces of potash, — the absorption-bulbs R are inserted. The safety-jar, which is the least important part of the whole apparatus, prevents the entrance of impure air at the end of a combustion, when a slight back-pressure is apt to take place. The bulbs R are made after the Geissler pattern, with the modern improvement of having the drying-tube joined to the bulbs in one piece. The bulbs are filled with a potash solution of the same strength as mentioned above; the drying-tube is filled with pieces of solid potash. Chloride of calcium cannot here be used effectively. When thus connected the apparatus should permit air being passed through it for hours without showing any increase of weight in the bulbs. Having ascertained this the apparatus can be safely used for many weeks, only renewing the pumice and the contents of the absorption-bulbs occasionally.

To carry out a combustion the platinum-funnel containing the carbon residue is dropped into the platinum-tube with the narrow end first. The above-mentioned platinum-gauze keeps the funnel in its proper position at the centre of the platinum-tube. The joint at the wider end of the platinum-tube is closed and the weighed absorption-bulbs inserted in their proper place. The flow of oxygen is started at the rate of about three bubbles a second, and a couple of good gas-burners lit under the tube and turned on full. The centre of the platinum-tube may be covered with a piece of some refractory material such as one-half of an "Erdmann's cylinder." After ten to twenty minutes the oxygen is shut off, the air put on and the gas turned down. When the tube has nearly cooled, the bulbs are taken out and weighed without delay. After accurately weighing the bulbs they may be immediately inserted in their place again and the next combustion proceeded with. It is quite unnecessary — in fact often inaccurate — to allow the bulbs to stand any length of time in the balance-case before weighing them or to wipe them off, as recommended by some writers. The amount of carbon is found from the increase of weight of bulbs (CO_2) by multiplying the said increase by $\frac{3}{11}$, or .2727.

b. **Determination of Carbon by Combustion with Chromic and Sulphuric Acids.** The carbon in the residue, after dissolving in the double chloride of copper and ammonia, may in many cases be determined with equal accuracy by oxidizing the same by means of chromic and sulphuric acids in a flask, arranged as for a sulphur determination (Fig. V.). From this flask the gas passes through a pair of sulphuric acid bottles arranged as in Fig. IV. C C_1, and from these to the chloride of calcium jar, absorption-bulbs, and safety-jar exactly as in the previous method. In obtaining the carbon residue a glass funnel-tube is used for filtering instead of a platinum-tube. The asbestos filter is supported by means of small pieces of broken glass. The suction-pump must be used as before. No drying of the residue is required. The funnel with its contents is placed at the bottom of the half-liter flask with long neck (Fig. V.). Through the funnel-tube twenty cubic centimetres of chromic acid solution (one weight of chromic acid in three weights of water) is poured; then one hundred and fifty cubic centimetres of concentrated sulphuric acid are run in (dilute acid must not be used). An excess of sulphuric acid is not objectionable. Care must be taken that the sulphuric acid be perfectly colorless and free from organic matter. Oxidation of the carbon sets in at once at a lively rate. After a short time heat is applied gradually until the solution has turned green, the evolution of oxygen slackened

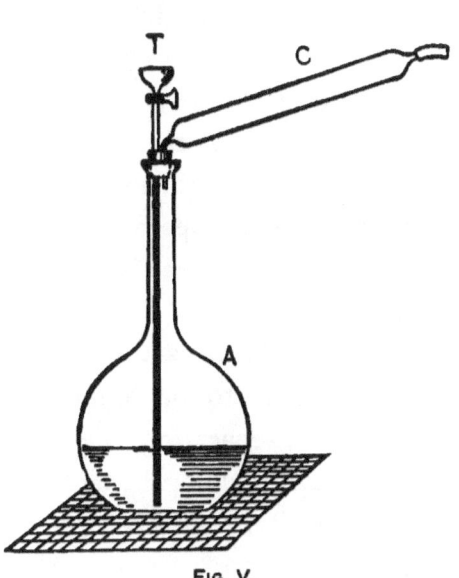

FIG. V.

and white sulphuric acid fumes have begun to appear. The heat is then removed and cold pure air pressed through the apparatus until the half-liter flask has somewhat cooled down. The absorption-bulbs may then be weighed and the next combustion proceeded with. The whole operation does not require more than one hour for completion. It does require, however, more attention than combustion with the platinum apparatus, and cannot compare with this latter in practical utility. The apparatus is cheap, which is a great item in its favor. Where large numbers of combustions have to be made, several apparatus and a large number of flasks with their respective funnels placed in them can be used.

c. **Other Combustion Methods for Carbon.** All those methods to be found in handbooks, which are based upon separating the carbon by means of ferric chloride, cupric chloride, cupric sulphate, mercuric chloride, electricity, etc., involve a loss of carbon as gaseous hydrocarbons and are entirely impracticable and unreliable. The use of the double chloride of copper and ammonia reduces this loss to a trifling minimum in most cases. An exception occurs in the treatment of highly carburetted ferromanganeses, for which *direct* combustion (see p. 50) should be used. Chromium steel also requires direct combustion, being difficult to dissolve in the double chloride. A requisite to success in determining carbon by direct combustion is to have the material in a fine state of division, which condition can be easily attained in the case of ferromanganeses and most chromium steels.

d. **The Determination of Carbon by Color.** The combustion methods enable us to determine with great accuracy the amount of carbon in irons and steels. By these methods we can prepare so-called standard steels, and upon the use of these is based the following color method: —

Two-tenths of a gram of each sample of drillings and of a standard steel are carefully weighed out and placed in test-tubes. It is advisable to use a standard in which the carbon is approximately the same as that in the samples. From a burette with glass stop-cock four cubic centimetres of nitric acid, specific gravity 1.20, are run into each test-tube. The tubes thus filled are immediately placed in a beaker, or tin can, containing cold water, in order to check the first violent action of the acid. The test-tubes should be about one hundred and thirty millimetres long and in inside diameter fifteen millimetres. The beaker, or can, containing the test-tubes, is first gradually warmed, which can be well done by placing it for a quarter of an hour on top of a boiling water-bath. This water-bath should consist of a tin or copper can of suitable size, provided with a perforated plate set at about two inches from the top edge. Through the perforations in this plate the test-tubes are put; thus, while resting on the bottom of the bath, they are steadied by the plate. The bath is then kept at a full boil for about half an hour by means of a powerful gas-burner (Fletcher's). The tubes are then taken out, cooled in cold water, and placed in a test-tube rack ready for testing. By the above operation the nitric acid solutions have assumed deeper or lighter colors, according to the quantity of carbon present.

For comparing these colors with the standard, we make use of three graduated tubes of thirty-five cubic centimetres capacity and thirteen millimetres internal diameter. These tubes should be of the whitest glass and exactly similar in all respects. They are graduated into tenths of cubic centimetres. In general working only two of the tubes are used, one for the standard, and one for the sample, the third tube being held in reserve for a second standard of different dilution and color, for very accurate work.

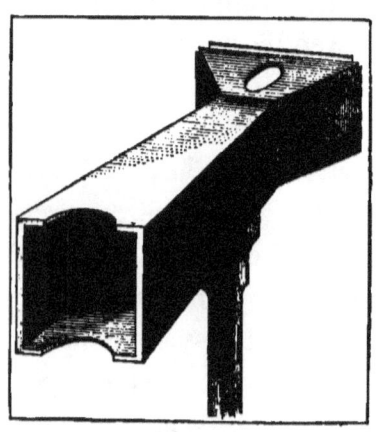

FIG. VI.

A good light is of importance; this is best obtained in a room with a single window facing north. A camera, made of blackened wood, as shown in Fig. VI., is very useful for securing a uniform light. One end of the camera, towards the light, is closed by means of a piece of ground glass. The tubes are inserted through a hole in the top of the camera and held up against the ground glass at a short distance from the same. The camera is of great use for making determinations at night.

To most eyes the left-hand tube appears a little the darker. The rule should therefore be to first hold the standard to the left and note when the test, on changing it from right to left after each successive dilution with water, shows a very faintly darker shade than the standard. After these general remarks the operation is easily understood.

Suppose the standard steel contain .84 per cent of carbon. We dilute it then with water to 16.8 cubic centimetres. Two-tenths of a gram of drillings having been used, each cubic centimetre will thus correspond to $\frac{.00168}{16.8}$ = .0001 gram of carbon; or, in other words, to one-tenth of one per cent of *one*-tenth of a gram of steel.

Suppose further that, on diluting the test to 14 cubic centimetres, we find the color of the test yet slightly darker than that of the standard; but, on diluting to 15 cubic centimetres, we find the test somewhat lighter than the standard. It is then evident that our test contains more than .70 per cent ($.1 \times \frac{14}{2}$) of carbon, but less than .75 per cent ($.1 \times \frac{15}{2}$). The steel can thus with safety be assumed to contain .72 or .73 per cent of carbon.

This is the only way of using the color-test that has proved thoroughly reliable in practical working, it being easily mastered by any boy of average intellect. To use a smaller quantity of drillings than two-tenths of a gram multiplies the errors from weighing and comparing of color and must be condemned. It is far easier to see when two shades of color differ slightly, than to catch the point where they are exactly alike. Unless the operator can do this, however, he cannot work successfully on less than two-tenths of a gram. On the other hand, it is not desirable to use more than two-tenths of a gram, as then the size of the measuring-tubes has to be increased, rendering the comparison of color more difficult.

The use of the color-test is limited by the fact that certain treatments which the steel may have undergone change the condition of the carbon; thus a steel shows less carbon by color when hardened than when unhardened, and more when annealed than when unannealed; it is therefore important to use standards that have undergone approximately the same physical treatment as the samples operated upon. When this is not practicable, combustion must be resorted to. There are many devices in the market for facilitating the color-test, called chromometers, etc. None of them offer any special advantage. The method of having standard solutions with so-called permanent colors, instead of dissolving a standard with every set of carbon determinations, is entirely untrustworthy. As to the quantity of acid used, it may be modified a little, so that somewhat more be used for very high carbons and less for low carbons. Care must be taken, however, that in *each set* of determinations the same quantity be used.

It is of course not necessary to dilute the standard so as to correspond exactly to .0001 gram per cubic centimetre. The above .84 standard diluted to 20 cubic centimetres would correspond to .000084 grams per cubic centimetre, and the carbon in our sample would have been found say between 17 cubic centimetres and 18 cubic centimetres, consequently between .71 per cent and .75 per cent ($\frac{17}{2} \times .084$ and $\frac{18}{2} \times .084$).

c. **Determination of Phosphorus.** Five grams of drillings are dissolved in a beaker (No. 5 Griffin's wide lipped) in nitric acid, 1.20 specific gravity. The solution is then evaporated with excess of strong hydrochloric acid by rapid boiling on a large iron plate heated by one or more Fletcher's solid flame burners. Water-baths and sand-baths are entirely unnecessary for the methods of analysis described in this book. The plate is so heated that the heat gradually decreases from the centre towards the edges. The hottest part ought to be rather above than below 300° C. The plate is a very simple, yet a most important and useful apparatus in an iron laboratory. The evaporation is continued on the hottest part of the plate until signs of spattering are noticed. The beaker, or beakers, are then moved to a less hot part of the plate. When the tendency to spatter has ceased the beakers are moved back to the hottest part of the plate and left there for at least half an hour. This heating is necessary in order to completely oxidize and decompose the last traces of iron phosphide, which otherwise would remain insoluble with the silica. The presence of hydrochloric acid lessens the tendency to spatter, which is always less in high carbon steels than in low carbon steels. The beakers are now slowly cooled and strong hydrochloric acid added in excess. The acid is at once brought to a boil, which effects a solution of the residue, and the boiling is continued until only a small bulk remains, with the silica sticking to the sides of the beaker. This boiling serves two purposes: firstly, to convert any pyrophosphoric acid ($H_4 P_2 O_7$), which may have been formed on the strong heating, into orthophosphoric acid ($H_3 PO_4$); secondly, to concentrate the solution, so as to render filtration easier and remove excess of hydrochloric acid, which would otherwise interfere with the precipitation of phosphorus by means of molybdic acid. Hot water is added to the concentrated solution and the silica filtered off on a four-inch Swedish filter-paper. It is well to collect the filtrate in a conical assay-flask. The silica must be well rubbed off from the sides of the beaker by means of a piece of rubber-tubing attached to the end of a glass rod. The washing of this silica is done by means of dilute hydrochloric acid and plenty of hot water. The silica thus obtained will in most cases burn perfectly white and can be used for the determination of silicon. The burnt silica contains .467, or $\frac{7}{15}$, of silicon. The phosphorus in the filtrate is precipitated as the yellow phosphomolybdate of ammonia. For this precipitation is used a solution of about one part by weight of molybdic acid in four weights of ammonia, .96 specific gravity, and fifteen parts of nitric acid, 1.20 specific gravity. The molybdic acid is first dissolved in the ammonia and this solution slowly poured into the nitric acid, which must be shaken constantly in order to prevent the separation of molybdic acid, which redissolves with difficulty. After a few days' standing the solution may be siphoned off clear. Fifty to one hundred cubic centimetres of this molybdic acid solution are used for each phosphorus determination. To precipitate the phosphomolybdate of ammonia in the filtrate from the silica, sufficient ammonia (.96 specific gravity) is added to nearly neutralize the solution. The fifty cubic centimetres of molybdic acid solution are then added and the flask well shaken. If the yellow precipitate is slow in coming down, a little more ammonia may be added. If too much ammonia is added, a little strong nitric acid must be introduced to redissolve the iron precipitate. As a rule the yellow precipitate comes down very quickly. By neutralizing the solution *before* adding the molybdic acid as described, the yellow precipitate becomes granular and easy to filter. When precipitated in any other way it has a great tendency to pass through and creep over the edges of the filter. The yellow precipitate is allowed to settle over night at about 40° C., or during a few hours at 80° C. After settling, the clear supernatant liquid is siphoned off and thrown away (or kept for reclaiming the molybdic acid). The yellow precipitate is filtered off on a four-inch Swedish filter of good quality and washed with copious quantities of the above-mentioned molybdic acid solution, diluted with an equal volume of water. About 300 cubic centimetres of washings are not too much to insure the complete removal of the last traces of iron. The yellow precipitate is then treated on the filter with a little hot ammonia, .96 specific gravity, and the filtrate allowed to run back into the assay-flask in which the precipitation was made. When all is dissolved, the ammoniacal solution is thrown on to the same filter again, but now to run into a 100 cubic centimetre beaker. The filter is washed with a little cold water, so that the bulk of the ammoniacal solution will be from 20 to 30 cubic centimetres. To the filtrate thus obtained are added about 3 cubic centimetres of hydrochloric acid, 1.12 specific gravity, and a few drops of ammonia to redissolve any yellow salt that may have separated. Then add 10 cubic centimetres of magnesia mixture and shake until the white crystalline phosphate of magnesia and ammonia appears $\left(\begin{smallmatrix} Mg \\ NH_4 \end{smallmatrix}\right\} PO_4 + 6\ H_2O)$.

About 6 cubic centimetres of ammonia, .96 specific gravity, are then added. If a good shaking now be applied the precipitate will be down completely in two hours; if, however, time is not available for shaking, the beaker must be left standing for at least six hours before filtering. With very vigorous shaking the precipitate can be brought down in half an hour. The white precipitate is filtered on to a two-inch Swedish filter-paper and washed with ammonia, .96 specific gravity, diluted with three times its volume of water. About 80 cubic centimetres of this mixture is sufficient for washing the precipitate. It is advisable not to use more than 80 cubic centimetres of wash-water, as the same has a slightly solvent action on the precipitate. The white precipitate must be rubbed loose from the sides of the beaker, like the silica, with rubber tubing on a glass rod.

The magnesia mixture is prepared by dissolving 110 grams of magnesium chloride together with 280 grams of ammonium chloride in 1,300 cubic centimetres of water and adding 700 cubic centimetres of ammonia, .96 specific gravity, to the solution. After a few days' standing the solution can be siphoned off clear from the sediment. The chloride of ammonium prevents the precipitation of magnesium hydrate. The following formulas show how this occurs:—

$$2\ MgCl_2 + 2\ NH_3 + 2\ H_2O = \left.\begin{matrix}Mg\\(NH_4)_2\end{matrix}\right\}Cl_4 + Mg\ H_2O_2,\ \text{and}$$

$$Mg\ H_2O_2 + 4\ NH_4Cl = \left.\begin{matrix}Mg\\(NH_4)_2\end{matrix}\right\}Cl_4 + 2\ NH_3 + 2\ H_2O.$$

The filter with the well-washed white precipitate is put, while still wet, into a weighed platinum crucible and ignited. The ignition is best conducted by keeping the crucible in an upright position and placing the lid in a slanting position in the same. At first only the top or oxidizing part of the burner-flame is allowed to strike the crucible. When all the filter-paper is burnt off more of the flame may be brought into contact with the crucible. Should some carbon still refuse to burn out, moisten with a little nitric acid and ignite again. The ignited precipitate is pyrophosphate of magnesia, $Mg_2 P_2 O_7$, containing very nearly 28 per cent of phosphorus. Care must be taken that the filter does not leave any appreciable amount of ash. This can be insured by washing every filter out with hydrochloric acid before using it.

The yellow precipitate, when dried at 95° to 140° C., contains 1.63 per cent of phosphorus and can be used for a direct phosphorus determination. It must then be washed with water containing 1 per cent by volume of nitric acid, 1.20 specific gravity, instead of with the dilute molybdic acid solution. After drying it is transferred from the filter, by shaking and brushing, into a weighed watch-glass, or some other suitable vessel, and weighed. When much phosphorus is present this method can be used with great accuracy, but when little the risk of loss in brushing off is too great. Weighed filters have then to be used. The magnesia method is, however, undoubtedly the best of the two methods in general working.

When precipitating phosphorus with the molybdic acid solution above described, it should be borne in mind that 100 cubic centimetres of the said solution are required for the complete precipitation of .1 gram of phosphoric anhydride ($P_2 O_5$), containing .044 gram of phosphorus. Ten cubic centimetres of the magnesia mixture are required for the same quantity of phosphorus.

f. **Determination of Manganese.** Five grams of sample are dissolved in about 150 cubic centimetres of dilute nitric acid, 1.20 specific gravity. The solution is then boiled on the plate (see p. 23) with the addition of strong nitric acid, 1.42 specific gravity, until the bulk is about 100 cubic centimetres. If the silicon in the sample is much higher than .2 per cent a clogging of the filter in the subsequent filtration may occur. In such cases it is therefore best to dissolve the sample in dilute hydrochloric acid and evaporate to gentle dryness. The dry mass is then dissolved in strong nitric acid and boiled to complete destruction of the hydrochloric acid. To the solution evaporated to 100 cubic centimetres small crystals of chlorate of potash are gradually added. Yellow and green fumes come off; and, after boiling has been continued for a while, all the manganese separates as dioxide (Mn O_2), insoluble in the strong nitric acid. The reactions may be considered to be,—

$$2 \text{ Mn O} + N_2 O_5 = 2 \text{ Mn } O_2 + N_2 O_3, \quad \text{and}$$
$$N_2 O_3 + Cl_2 O_5 = N_2 O_5 + Cl_2 O_3.$$

The chlorous acid gas ($Cl_2 O_3$) is very explosive. By using a smaller bulk than 100 cubic centimetres violent explosions may occur, throwing the lid off the beaker. On the other hand it is not good to use a much larger bulk than 100 cubic centimetres, as the solution foams rather violently on addition of the chlorate of potash. Most likely the reactions are far more complicated than as above given, the higher oxides of manganese first forming and the dioxide of manganese separating on boiling.

After boiling the precipitated manganese dioxide for a few minutes some cold strong nitric acid is added and the precipitate filtered off on an *ordinary* asbestos filter in a glass tube by means of the suction-pump. Care must be taken that the nitric acid here used contain no nitrous acid ($N_2 O_3$), the presence of which would cause a reduction of the manganese dioxide to lower oxides. A yellow color indicates the presence of nitrous acid in the nitric acid. Bottles that have been contaminated with nitrous acid and nitric oxide gas ($N_2 O_4$) are easily cleaned by shaking them with water, which dissolves said impurities. The manganese dioxide is washed on the filter twice with cold strong nitric acid and four times with cold water. After washing, the precipitate together with the asbestos is blown back into the beaker in which the precipitation was originally made. The manganese may now be determined either gravimetrically or volumetrically with equal accuracy. The latter method, on account of its rapidity, is always preferable where large numbers of determinations are constantly made.

(a) *Gravimetric Determination.* Boil the precipitated dioxide with hydrochloric acid until all chlorine is driven off, the manganese being thereby converted into the proto-salt (Mn O). Dilute with water and filter off the asbestos. Add a little concentrated acetic acid to the hot filtrate and neutralize with ammonia, .88 specific gravity, until the solution barely smells of acetic acid. Boil, and allow the precipitated basic acetate of iron (a little iron always accompanies the $Mn\ O_2$) to settle. If the smell of acetic acid now seem rather strong add a little dilute ammonia. Then filter and wash well with hot water. If the precipitated oxide of iron looks reddish all over, it may be considered as free from manganese; but if brownish flakes appear floating on the surface during the filtering, the presence of manganese oxides is indicated, and it becomes necessary to redissolve the precipitate by boiling with hydrochloric acid and to repeat the basic acetate precipitation as above directed in order to insure the complete separation of the iron from the manganese. To the filtrate, or filtrates, from the above basic acetate precipitation an excess of the strong ammonia is added and then, with vigorous stirring, a few cubic centimetres of bromine, by which the hydrates of the higher oxides of manganese are precipitated. Heat gradually to boiling, allow to settle, filter, ignite in a platinum crucible and weigh as $Mn_3\ O_4$, containing 72.08 per cent of Mn. The ignition must be either very long and protracted over an ordinary Bunsen's burner or short over the blast-lamp. The washing of the manganese precipitate is easily accomplished by means of hot water, there being only volatile ammoniacal salts present and no fixed alkali.

(β) *Volumetric Determination.* The volumetric determination of the manganese in the above-described dioxide precipitate, in which all the Mn exists as dioxide, consists in general in dissolving the dioxide in a measured amount of a standard acid solution of ferrous sulphate, by which the Mn O_2 is at once reduced to Mn O, an excess of ferrous sulphate remaining. The amount of this excess is then determined with a standard solution of bichromate of potash, from which data the amount of manganese dioxide dissolved in and reduced by the acid ferrous sulphate can be estimated. For the method as above mentioned is required a standard solution of ferrous sulphate, a standard solution of bichromate of potash and a solution of ferricyanide of potash (K_6 CN_{12} Fe_2). The last-named solution is prepared separately for every set of determinations. The ferrous sulphate solution is prepared by dissolving 20 grams of ferrous sulphate (Fe SO_4 + 7 H_2 O, containing one-fifth part of iron) in 1,600 cubic centimetres of water and 400 cubic centimetres of sulphuric acid of about 1.5 specific gravity. This solution, when kept in darkened two-liter bottles with ground stoppers, will not perceptibly change in strength (i.e., will not be oxidized) even when kept for a long time. The bichromate solution is prepared by dissolving about 10 grams of the salt in 1 liter of water. One cubic centimetre of this solution will very nearly oxidize such an amount of the ferrous sulphate to ferric sulphate as would contain .011 of a gram of iron, as per the following formulas of combining proportions:—

$$6\,FeSO_4 + K_2Cr_2O_7 + 7\,H_2SO_4 = 3\,Fe_2\,3\,SO_4 + K_2SO_4 + Cr_2\,3\,SO_4 + 7\,H_2O.$$

One hundred cubic centimetres of the ferrous sulphate solution, of the strength as above directed, will correspond to 18.1 cubic centimetres of the above bichromate solution. The ferrous sulphate solution must be checked for strength with every set of determinations, if the intervals of time between such determinations is considerable. The standardizing of the bichromate solution may be effected by making therewith three separate determinations of iron in three carefully weighed amounts — from one to two grams each — of ammonio-ferro-sulphate $\left(\begin{smallmatrix}(NH_4)_2\\Fe\end{smallmatrix}\right\}2\,SO_4 + 6\,H_2O\right)$, which salt can always be obtained pure, and then contains one-seventh part of iron. Pure iron wire may also be used for standardizing the bichromate. The bichromate solution is made up in large quantities, as it will retain its strength unchanged for years. It should be kept in a large bottle on a high shelf, so arranged that the solution can be conveniently run into a 100 cubic centimetre Mohr's burette, with Erdmann's float, through a tube attached to the lower part of the bottle. A porcelain dish of at least one liter capacity is placed under the burette. This dish receives the solution to be titrated. In a small beaker some crystals of ferricyanide of potash are dissolved in water. There is no need of weighing out an exact quantity; but enough should be added to give the solution a *bright* yellow color, not a greenish or pale yellow tint. Drops of this solution are placed in the cavities of a porcelain test-plate, such as is used by artists for water-colors. By conveying drops of the solution in the dish to be titrated on a glass-rod, and mixing them with the yellow drops of the ferricyanide, we notice how the resulting blue color, on addition of bichromate to the solution in the dish, becomes first lighter and then greenish; and finally, after the addition of a last one-tenth of a cubic centimetre of bichromate, how the drops show a decided brownish tint. This brownish tint indicates that the last trace of protoxide of iron has been converted into peroxide.

The determination of the manganese is now easily understood. Add from a pipette 100 cubic centimetres of the standard ferrous sulphate solution to the MnO_2 precipitate, which has been blown back into the beaker as described, and shake well until all the manganese is decomposed and dissolved. Then determine by titration in the dish how much iron has been left unoxidized by the manganese dioxide. The reaction is —

$$MnO_2 + 2\,FeO = Fe_2O_3 + MnO.$$

In the above formula one part of iron corresponds to .491 part of manganese. A table may thus be prepared and can be pasted on to the bottle containing the bichromate solution as follows: —

ANALYSIS OF WROUGHT IRON AND STEELS.   35

1 cubic centimetre bichromate = .011 gram Fe.
1 cubic centimetre bichromate = .0054 gram Mn.

Or, when 5 grams of sample have been used, —

$\frac{.0054}{5}$ = .00108 grams Mn
= .108 per cent of Mn in sample.

Suppose 100 cubic centimetres of the ferrous sulphate correspond to 18.1 cubic centimetres of the bichromate and, further, that 15.1 cubic centimetres of the latter are found, by titration, to be required for oxidizing the ferrous sulphate left unoxidized by the manganese dioxide, then we have 18.1 — 15.1 = 3 cubic centimetres of bichromate, corresponding to the iron that has been oxidized in the ferrous sulphate by the manganese dioxide. Having used five grams of sample, we find direct from the above table 3 × .108 = .324 per cent of manganese. The top of the burette should be kept closed by a piece of cotton, to prevent the entrance of organic matter which would cause a slow decomposition of the bichromate when exposed to strong sunlight. The burette must always be kept filled to the top.

g. **Determination of Silicon.** The silica obtained in the phosphorus determination (see p. 24) is apt to contain some impurities. It is therefore advisable to use the following special method which has a more general application and which gives a pure silica in the largest number of cases.

Five grams of sample are dissolved in such a quantity

of dilute sulphuric acid that on each gram of drillings come 2 cubic centimetres of the strong acid. When all is dissolved strong nitric acid is cautiously added until no more effervescence occurs. The solution is then boiled down on the plate until white fumes appear; the dry mass is moistened with a little hydrochloric acid and then dissolved in a small excess of boiling water. When solution has been effected, the silica is filtered off, washed with dilute hydrochloric acid and hot water, ignited and weighed (see p. 24).

h. **Determination of Sulphur.** (a) *Bromine Method.* Ten grams of sample are weighed out and put into the half-liter flask A (Fig. VII.) with long neck. The flask is connected with the absorption-bulbs B containing hydrochloric acid, 1.12 specific gravity, and about 5 cubic centimetres of bromine. The wide tube C causes the vapor to condense and flow back into the flask A during boiling. The bulbs B connect with a long glass tube by which the bromine fumes can be carried off through a hole in a window, or, better still, through a flue with strong draught. The connections being made, 200 cubic centimetres of boiling water are run in through the thistle-tube T. The air is thus completely driven out of the flask. Two hundred cubic centimetres of strong hydrochloric acid are then added. When the gas begins to run rather slowly through B heat is applied until boiling gradually ensues. When the steel is completely dissolved the apparatus is disconnected and the contents of the bulbs B rinsed out into a beaker of 100 cubic centimetres capacity, into which a few cubic centimetres of a concentrated solution of chloride of barium (100 grams of Ba Cl$_2$ in 1 liter of water) have been previously introduced. Heat is then applied to the beaker on the hot iron plate until the bromine is completely driven off and the sulphate

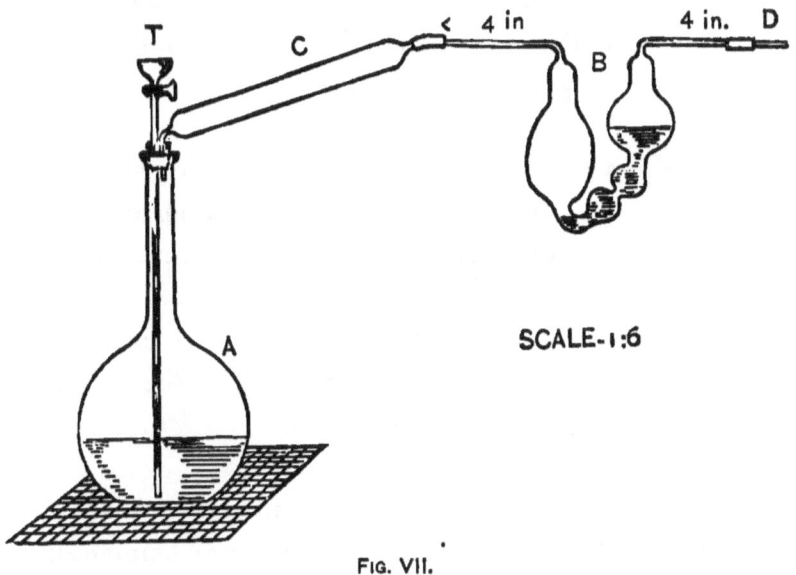

Fig. VII.

of barium has settled nicely to the bottom. The barium sulphate is then filtered off on a small double filter, washed with hot water and finally ignited and weighed. The filter should always be put into the crucible whilst still wet. The ignited barium sulphate contains 13.72 per cent of sulphur. A perfectly pure barium sulphate is obtained by this method, there being no bases present by which it can be contaminated. During the passage of the gas through the bulbs B, oily drops of propyl-bromide, etc., are formed, which, however, disappear on heating. The reaction, by means of which the sulphur is retained in the bromine solution, is as follows: $H_2 S + 4 H_2 O + 8 Br = H_2 SO_4 + 8 H Br$. For making many determinations in succession it is convenient to have a large number of bulbs filled and suspended in a box of suitable form, as well as many flasks. The flasks must be perfectly dry before putting the drillings into the same.

(b) "*Aqua Regia*" *Method.* If the iron or steel contain much copper, — say more than one-fourth of one per cent, — or other elements that are precipitated by $H_2 S$ in acid solution, the bromine method is apt to give too low results. If the samples be very coarse the solution in H Cl proceeds too slowly and the bromine method becomes difficult to manage. In these cases the following method can best be used.

Five grams of sample are treated in exactly the same way as for phosphorus determination, care being taken to have all reagents perfectly free from sulphur and also taking the precaution to add a little sodium carbonate before drying on the iron plate. If this precaution be omitted some sulphuric acid may be lost through volatilization. To the filtrate from the silica — which should amount to at least 300 cubic centimetres — a few cubic centimetres of $Ba Cl_2$ solution are added. After standing one night at the temperature of the room the sulphate of barium is completely precipitated and can be filtered off, washed thoroughly in the beaker with strong H Cl and on the filter with dilute H Cl and hot water, ignited (compare p. 36), and weighed.

i. **Determination of Copper.** Five grams of sample are dissolved in a mixture of 10 cubic centimetres of strong sulphuric acid and 150 cubic centimetres of water. Heat to full boiling and add, with constant stirring, 2 cubic centimetres of a concentrated solution of the thiosulphate — formerly known as "hyposulphite" — of soda ($Na_2 S_2 O_3 + 5 H_2 O$). The thiosulphate solution is obtained of proper strength by dissolving eighteen parts of the salt in twenty parts of water. The copper is precipitated as subsulphide ($Cu_2 S$) together with free sulphur. This subsulphide does not oxidize in the air during washing, etc., which is of great advantage. It is filtered off and washed somewhat with hot water. The whole precipitate, including carbon residue, free sulphur, iron, etc., is washed back into the beaker in which the precipitation was made, dissolved by boiling with some *aqua regia* and evaporated with sulphuric acid to complete expulsion of H Cl and H N O_3. Dilute with cold water and precipitate the iron present with excess of strong ammonia at a boiling heat. Filter off the precipitate and wash with ammoniacal water. If any appreciable amount of copper is present, the filtrate will appear blue or bluish-green. Acidify with sulphuric acid in *slight* excess, heat to boiling and precipitate the copper, now pure, with a few drops of thiosulphate. Filter off, put the precipitate whilst wet in a weighed *porcelain* crucible and ignite strongly. Weigh as cupric oxide (Cu O), containing 79.8 per cent of copper.

In the method above described the iron was originally present as Fe O. If, however, copper is to be determined in a solution containing Fe_2O_3, much more thiosulphate solution is required, as the Cu_2S does not come down until all the iron has been completely reduced, as per formula: —

$$Fe_2\ 3\ SO_4 + 2\ Na_2S_2O_3 = 2\ Fe\ SO_4 + \underset{\text{(Sodium tetrathionate.)}}{Na_2S_4O_6} + Na_2SO_4.$$

Thus if in a solution of 5 grams of borings all the iron exists as Fe_2O_3, about 30 cubic centimetres of the thiosulphate would be required. In such cases it is easy to notice, with sufficient accuracy for the purpose, the point when enough thiosulphate has been added. As long as there is any peroxide left, a black color appears and again disappears after each addition of thiosulphate. Finally the color of the solution changes to a light green. This is the point when enough has been added. A little sulphur then begins to separate and, if any copper be present, a black precipitate soon appears on continued boiling. Hydrochloric acid must be absent in this operation as it seems to be impossible to effect complete precipitation of Cu in its presence. Nitric acid retards the precipitation in the same way as Fe_2O_3, inasmuch as the same must be reduced before the Cu is precipitated. If great excess of sulphuric acid and thiosulphate be used the copper will not be completely precipitated. Two cubic centimetres of sulphuric acid solution, 1.83 specific gravity, are required to decompose 5 cubic centimetres of the above thiosulphate solution and 1.05 cubic centimetres of the same strong sulphuric acid are required to dissolve 1 gram of iron. One-tenth gram of copper, contained in solution as sulphate and dissolved in 50 cubic centimetres of water, is completely precipitated by 1 cubic centimetre of the thiosulphate solution.

The reaction according to which the Cu_2S is obtained is unknown. It is considered that at first thiosulphate of copper is formed which, on boiling, decomposes into Cu_2S and into a sulphuric acid of unknown composition.

It must be remembered that many other metals — such as arsenic, antimony, lead, tin, bismuth, etc. — are also precipitated by the thiosulphate as sulphides. In wrought iron and steel, however, these metals do not often occur in sufficient quantity to cause any error.

j. **Determination of Slag and Oxide of Iron.** Slag and oxides of iron usually occur in wrought iron in considerable quantities. Steels are, as a rule, comparatively free from the same. When, however, these impurities do occur in steels in unusual quantities, the quality of the metal is seriously affected thereby and their determination becomes important. It has been found very difficult to make such determinations with accuracy and, although it is believed that the following method is superior to those usually proposed, it must be borne in mind that it does not give results which can be compared in accuracy with those obtained by the methods already given for the determination of other impurities.

Five grams of drillings are treated with 250 cubic centimetres of a ferric chloride solution containing about 1 part of crystals of ferric chloride to 1.5 part of water. Care must be taken to use chemically pure ferric chloride, and the crystals should smell somewhat of chlorine. The water should be previously freed from air by boiling and subsequent cooling to the ordinary temperature. A flask similar to those used in carbon determination (p. 7) is well suited for the decomposition, which takes place as per formula: —

$$Fe_2 Cl_6 + Fe = 3 Fe Cl_2.$$

The residue contains all the slags and oxides of iron, some free silica and silicon and, further, various compounds of iron with other bodies, such as, principally, phosphorus, sulphur, arsenic and carbon. Nearly all the carbon is left with the residue.

It should be borne in mind that wrought irons and steels are produced by oxidizing processes, whilst pig-iron is produced by a reducing process. The former, during their manufacture, are, therefore, apt to lose various ingredients that have greater affinity for oxygen than has iron, — such as silicon, manganese, titanium, chromium, vanadium etc., — whilst the latter may take up any number of such elements. Copper, cobalt and nickel are more easily reduced from their oxygen compounds than iron, and therefore generally remain with the iron through all stages of manufacture. From the above it follows that the residue obtained by the treatment of wrought iron and steel with ferric chloride contains, as a rule, fewer compounds of iron with other elements than when pig-iron is so treated. Sometimes, however, such compounds do occur. The author found lately in some steel-blooms made by the "basic" process a black compound, when searching for silicon according to the methods described. This black compound resisted all acids and even the strong heating in the determination of phosphorus failed to decompose it. Fusion with alkaline carbonates was necessary to effect decomposition. It was present in the steel to the amount of .07 per cent, and consisted of iron and vanadium with indications of the presence of another not yet identified element. Thus it appears that certain compounds of iron may be formed in the pig-iron which resist all subsequent oxidizing influences, and are found in the refined products; i. e., wrought iron and steel. Such compounds are apt to be discovered in the ordinary silicon determinations and need not be suspected in the ferric chloride residue, in the case of wrought iron and steel, unless thus detected beforehand. The compound in question may also have been introduced into the steel with the final additions of spiegel, etc.

To separate the slag and oxides of iron from the ferric chloride residue it is necessary to treat the same with a hot alkali; by this means the silica is removed, any free silicon being oxidized to silica with the evolution of hydrogen gas. The residue must then be ignited in a current of dry chlorine to remove iron phosphides, arsenides etc.

In the case of wrought irons, which usually contain considerable slag and oxides of iron, a sufficient idea of the amount of these impurities can be gained by simply igniting and weighing the ferric chloride residue after washing thoroughly with hot alkali and cold water free from air.

For steels this treatment will give an approximate idea of the amount of slag and oxide of iron present; but it must be borne in mind that the residue also contains the phosphides, sulphides, etc. of iron and possibly other iron compounds. The method is, therefore, in this case, only suitable as a comparative test at the same steel-works where the chemical composition is kept tolerably uniform.

k. **Determination of Arsenic.** Ten grams of drillings are dissolved in dilute nitric acid in a beaker and evaporated with sulphuric acid to complete expulsion of the nitric acid. To facilitate the removal of the nitric acid the mass should be stirred with a glass rod. The heating is continued until the mass has become so dry that it can be shaken down into a half-liter flask. The mass is mixed in the half-liter flask with about 15 grams of powdered ferrous sulphate. Into the neck of the flask is inserted a rubber-stopper containing a funnel-tube and a bent glass tube from which a 50 cubic centimetre pipette is suspended by means of a rubber-tube. The lower end of the pipette dips down into a beaker filled with water. Through the funnel-tube about 100 cubic centimetres of strong hydrochloric acid are run in, and the flask gently heated. The ferrous sulphate reduces the arsenic acid to arsenious acid and the arsenic distils over into the beaker as arsenic chloride ($AsCl_3$). After about half an hour, when the pipette has become warm, all the arsenic present has been separated. By saturating the solution in the beaker with sulphuretted hydrogen at a temperature of about 70° C. and driving off the excess of H_2S with a current of carbonic acid gas passed through the solution, we obtain a yellow precipitate of As_2S_3. This is filtered off rapidly and dissolved in *aqua regia*. After boiling off excess of acid the arsenic is precipitated in a 100 cubic centimetre beaker as $\left({Mg \atop NH_4} \right\} As\, O_4 + 6\, H_2 O)$ by magnesia mixture, in exactly the same manner as is phosphorus. This precipitate, when gently ignited in a slow current of oxygen, yields $Mg_2As_2O_7$, containing 48.42 per cent of arsenic.

l. **Determination of Titanium.** The solution of ten grams in HCl from the bromine method for sulphur may be used. Add to this solution a small quantity of bromine, so as to oxidize only a small part of the iron. By now making a basic acetate separation of that portion of the iron thus oxidized we obtain all the titanium with the iron precipitate. Redissolve the basic acetate in H Cl and precipitate with ammonia. Ignite the precipitate in a platinum crucible, evaporate with hydrofluoric acid and ignite again, after adding a little sulphuric acid. Treat the residue with strong hydrochloric acid. This leaves any titanic acid ($Ti\ O_2$) undissolved. The latter may then be ignited and weighed as $Ti\ O_2$, containing 60 per cent of titanium.

m. **Tracing of Vanadium.** Vanadium, when present, usually accompanies the silica. By removing the latter with hydrofluoric acid and igniting the residue the vanadic acid fuses and, on cooling, forms a beautiful crimson crystalline mass. By fusing silica containing vanadium with sodium carbonate a greenish yellowish mass is obtained which dissolves in water, leaving some oxide of iron. In the solution the reactions for vanadium may be obtained.

n. **Determination of Chromium.** The following method is the one most commonly used. It should here be pointed out, however, that the filtrate from the $Mn\ O_2$ precipitate in the nitric acid and chlorate of potash method for determining manganese (see p. 29) contains as chromic acid most and possibly all the chromium present in the metal. This could probably be made the basis for a rapid method of determining chromium in irons and steels.

ANALYSIS OF WROUGHT IRON AND STEELS. 47

One to five grams of sample are dissolved by boiling with hydrochloric acid, 1.12 specific gravity, in a flask of half a liter capacity. The neck of the flask should be closed with a rubber-stopper. Through a hole in this rubber-stopper is inserted a glass tube, attached to which is a piece of rubber-tubing. This rubber-tubing is closed at the other end by means of a piece of glass-rod and provided with a longitudinal slit through which the steam can escape during boiling, whilst no air can enter. Oxidation is, by this arrangement, almost completely prevented.

When all is dissolved, the flask is nearly filled with cold water and the solution neutralized with powdered carbonate of baryta in excess. If much free acid be present, some carbonate of soda should be added *before* the carbonate of baryta. Cork the flask tightly and allow to stand for twenty-four hours at ordinary temperature with occasional shaking. $Cr_2 O_3$ and a little $Fe_2 O_3$ are hereby precipitated, whilst all $Fe Cl_2$, $Mn Cl_2$ etc. remain in the solution. Filter off the precipitate, together with the excess of carbonate of baryta, wash with hot water and redissolve in H Cl by boiling. To the boiling solution thus obtained add ammonia, and boil off the excess of the latter. The $Cr_2 O_3$ and $Fe_2 O_3$ are hereby precipitated, whilst all the barium remains in solution. Filter off the precipitate, wash well with hot water, dry, ignite and fuse in a platinum crucible with carbonate of soda and a little nitrate of soda. Extract the fused mass with hot water and filter off the residual iron oxide. The filtrate contains all the chromium as the yellow sodium chromate. Acidify the filtrate with H Cl, add a little sodium sulphite and heat to boiling. The Cr O_3 is hereby instantly reduced to $Cr_2 O_3$. To the boiling green solution thus obtained add ammonia in slight excess. The color of the solution changes to violet at first; but, on continued boiling, the green chromic oxide is precipitated. When the smell of ammonia has almost completely disappeared the chromic oxide is filtered off, ignited in a platinum crucible and weighed as $Cr_2 O_3$, containing 68.62 per cent of chromium.

If the chromium steel contains silicon, it will be found as sodium silicate together with the sodium chromate and should be separated by evaporation in the usual way before precipitating the chromic oxide with ammonia.

o. **Determination of Tungsten.** Three to five grams of sample are treated in exactly the same way as in a phosphorus determination with the exception that the heating need not be so strong and protracted. Presence of tungsten is at once indicated by the yellow color of the tungstic acid (W O_3) which all separates with the silica.

The silica and tungstic acid are filtered off and washed with water, to which a little H Cl has been added to prevent the W O_3 from passing through the filter. The tungstic acid is then dissolved on the filter in hot ammonia, and is thus separated from the silica. The filtrate is concentrated so as to allow of its being poured into a weighed platinum crucible, in which it is evaporated to dryness, ignited, and weighed as W O$_3$, containing 79.3 per cent of tungsten.

B.—*Analysis of Pig-iron*.

The analysis of pig-iron is, with a few modifications, similar to that of wrought irons and steels.

a. **Determination of Graphite.** To determine separately the graphite and the combined carbon, we first determine the total carbon by the double chloride method (p. 7). The graphite is determined by dissolving five grams of drillings in dilute hydrochloric acid and filtering off the residue in the platinum funnel. This residue must now be freed from all nongraphitic carbon. This is done by washing first with hot water and hot ammonia and then with alcohol, ether, and, finally, again with water. A brisk evolution of hydrogen gas takes place when washing with hot ammonia, owing to the oxidation of silicon. The graphite is then burnt out in the platinum tube as usual and deducted from the total carbon. This gives us both the graphite and the combined carbon.

b. **Determination of Silicon.** For the determination of silicon the sulphuric acid method must be used exclusively. The silica cannot be obtained pure in the determination of phosphorus, as in the case of wrought iron and steel, whilst the sulphuric acid method yields a perfectly white silica with possibly some vanadic acid adhering to it. The vanadic acid imparts a yellowish brown color to the silica. It may, to some extent, be removed by washing with ammonia or with hot hydrochloric acid.

In the case of very silicious pig-irons one gram is sufficient for a determination of silicon.

c. **Determination of Sulphur.** The bromine method is in most cases applicable, but it is advisable always to use the *aqua regia* method as well, pig-irons being apt to contain more metals that are precipitated by H_2S than do wrought iron and steel. It is to be remembered, however, that it is extremely difficult to obtain the barium sulphate perfectly pure when using the *aqua regia* method for pig-iron.

d. **Slag** generally occurs in pig-iron and is found in the ferric chloride residue (p. 41). This slag carries lime, magnesia, alumina, silica etc., as do all blast-furnace slags. The ferric chloride residue may be analyzed completely according to directions given below for iron ores.

C.—*Analysis of Spiegel and Ferromanganese.*

The alloys of iron and manganese that are now manufactured may contain a proportion of manganese varying from a few per cent to 80 or 90 per cent. The determination of the manganese in these alloys is a most important one. If, as for most practical purposes, it be sufficient to determine the manganese to within one-half per cent, the simplest way is to determine the iron and to ascertain the manganese by difference, allowing for carbon, silicon, etc., according to the following table: —

When the iron contained is
>   Less than 20 per cent, deduct 7.5 per cent.
>   Between 20 and 45 per cent, deduct 6.5 per cent.
>   Between 45 and 65 per cent, deduct 6.0 per cent.
>   65 per cent and upwards, deduct 5.5 per cent.

This indirect method is only applicable to manganese alloys manufactured in the blast-furnace from manganiferous ores. Many manganese alloys are manufactured in crucibles and for these the table cannot be used. The iron is determined by dissolving .5 to 2 grams of the sample in dilute sulphuric acid and titrating with the bichromate of potash (see p. 33). For alloys very rich in manganese, however, this bichromate titration is not suitable on account of the brown precipitate which the ferricyanide of potash gives with the manganese. This brown precipitate obscures the final change of color. As a rule, this indirect method may be well used for alloys containing about 50 per cent of Mn and below.

The same methods as are used in the case of wrought iron and steel may also be here used for the determination of manganese, with the following modifications: The quantity of sample weighed out must be proportioned according to the amount of manganese present, so that a smaller quantity be used for a larger percentage of manganese and vice versa. If the amount of manganese exceeds the amount of iron present, about one gram of pure iron wire must be added. A still larger amount of wire is not objectionable; in fact, it ought always to be added when using the nitric acid and chlorate of potash method for alloys rich in manganese. If an excess of iron be not present, the manganese cannot be completely precipitated as MnO_2. This is a most important fact, although we have no satisfactory explanation of the cause thereof. A safe rule to follow is to combine the manganese alloy to be operated upon with such a proportion of iron wire that a total sample of five grams will not contain more than two per cent of manganese.

In the case of alloys very rich in manganese, say 75 to 90 per cent, it is necessary to use such a small quantity of the material to be analyzed that the nitric acid and chlorate of potash method with subsequent titration is not to be recommended and the following simple method is preferred: Weigh out one-tenth gram, and treat as for a phosphorus determination. Make two successive basic acetate precipitations (p. 30), without filtering off the silica, and determine the manganese in the filtrates with bromine and ammonia as usual. Sometimes even three basic acetates are necessary.

Carbon can be determined by combustion, as in wrought iron and steel, except in the case of the higher manganese alloys. When these are treated with the double chloride, a portion of the carbon is apt to escape in the form of gaseous hydrocarbons. In such cases five grams of the finely pulverized material should be mixed with pure cupric oxide in a platinum-funnel, and burned in a stream of oxygen for one hour or longer. The burning and the weighing of the bulbs should be repeated once or twice, to make sure that all the carbon is burnt out. Some chromium steels must be treated in the same way, as they are not completely decomposed by the double chloride.

Otherwise the analysis of the manganese alloys is similar to the analysis of wrought iron and steel. *Sulphur*, however, rarely occurs except in the smallest traces; in fact, it is greatly to be suspected that, whenever sulphur is found, it is due to impure reagents or incorrect manipulations.

When applying the method for *copper* determination, it must be borne in mind that we have to purify the first precipitate of copper subsulphide not only from iron, but also from manganese, and that, therefore, some bromine must be used along with the ammonia.

D.—*Analysis of Silicon-iron, and Silicon-manganese-iron.*

There is nothing specially to be remarked concerning the analysis of these alloys, except that the silica, even when obtained by the sulphuric acid method, is apt to contain many impurities, particularly vanadic acid. The silica should, therefore, after weighing it in the impure state, be volatilized with hydrofluoric acid, and the residue ignited, weighed, and the weight deducted from the total weight. A little sulphuric acid must be used, together with the hydrofluoric acid, to prevent any partial volatilization of the residue.

It is natural that alloys like the ones in question should contain more rare impurities even than pig-iron, the reducing action of the blast-furnace required in their manufacture being much greater than is the case with the latter.

CHAPTER III.

DETERMINATION OF THE MOST IMPORTANT INGREDIENTS IN IRON ORES, SLAGS, LIMESTONES, FUEL, ETC.

A.—*Analysis of Iron Ores.*

By iron ores we mean such minerals as are used for the manufacture of iron. Iron ores consist of the oxides, the hydrated oxides, and the carbonates of iron, contaminated with varying amounts of rocky or earthy matter. The residues after burning iron pyrites, puddel-slags, etc., are occasionally used as iron ores.

Manganese ores and manganiferous iron ores (carbonates of iron and manganese) also play an important part in modern iron and steel manufacture.

In practical work it is often only necessary to determine the most important elements entering into the composition of iron ores, such as iron, phosphorus, manganese, silica, insoluble residue, etc. As the direct methods for determining these substances are very similar to the corresponding methods employed in the case of wrought iron and steel, they are here first given; and the directions for making a complete analysis of iron ores, considered as minerals, will follow (see p. 61).

a. **Determination of the Total Iron.** Dissolve about one gram of finely-ground ore in dilute sulphuric*

---

* Or in hydrochloric acid, with subsequent addition of sulphuric acid and expulsion of the hydrochloric acid.

acid by boiling. When dissolved as far as possible, add some water and sulphite of soda, powdered, in such excess, that the solution assumes a dark-red color. After some heating, add more sulphuric acid, and boil until all smell of sulphurous acid has disappeared. The solution is now colorless, or perhaps slightly brownish, owing to organic matter, and all the iron is in the state of Fe O. We may thus determine the total iron by titrating with the standard bichromate of potash, as described above. This method will give the iron within about .5 per cent. This is for most practical purposes sufficiently accurate. A little iron nearly always remains in the residue, a little sulphurous acid remains in the solution, and a little Fe_2O_3 remains unreduced; all these circumstances interfere with obtaining a closer result.

Thiosulphate of soda cannot be used instead of the sulphite, owing to the formation of tetrathionic acid, which has a reducing action upon the bichromate.

Zinc also cannot conveniently be used as a reducing agent, owing to the brown precipitate which it forms with ferricyanide of potash. Zinc can be used with a standard oxidizing solution of permanganate of potash; but the disadvantages of the permanganate solution are so great, that it has almost everywhere been abandoned in favor of the bichromate.

It is better to use sulphuric acid than hydrochloric acid, for the reason that the final reaction with the ferricyanide is sharper.

---

* This can be obviated by using some H Cl with the solution of ferricyanide.

b. **Determination of the Iron present as Ferric Oxide (Fe_2O_3).** This method is carried out by dissolving 1.5 grams of ore in hydrochloric acid, and titrating with a standard solution of protochloride of tin. Use is made of a burette and a porcelain dish exactly similar to those used in the bichromate method. The standard protochloride of tin solution is also kept in a jar similar to that used for the bichromate; but, owing to the rapid absorption of oxygen from the air by the protochloride of tin, it should be covered with a layer of petroleum. When thus protected, it keeps for a very long time at its original strength. From the burette the standard solution is run into the ore solution, which should be boiling hot, and have a large excess of hydrochloric acid present, say about 25 cubic centimetres of H Cl, 1.12 specific gravity, for every one-tenth of a gram of iron. It is easy to observe the final reaction when the solution turns colorless. The reaction is —

$$Fe_2Cl_6 + Sn\,Cl_2 = 2\,Fe\,Cl_2 + Sn\,Cl_4.$$

The protochloride of tin solution is prepared by dissolving metallic tin in hydrochloric acid, 1.19 specific gravity, until evolution of gas has ceased, pouring off from excess of tin and diluting with H Cl, 1.12 specific gravity, to ten times the volume of the concentrated solution. The salt $Sn\,Cl_2$ can be used instead of metallic tin. The $Sn\,Cl_2$ solution is standardized by means of a ferric-chloride solution of known strength.

The Fe_2Cl_6 solution should have 10 grams of iron in one liter, and may be prepared either from iron wire or from ferric oxide. 10.04 grams of iron wire, or 14.2857 grams of Fe_2O_3, give 10 grams of iron. The iron wire is dissolved in H Cl and boiled with a little K Cl O_3 until all Cl is driven off. The ferric oxide is simply dissolved in strong H Cl. Fifty cubic centimetres of this iron solution contain .5 gram of iron, and about 25 cubic centimetres of the above tin solution will be required for this quantity. Three or four check titrations should be made, and the tin solution re-standardized frequently. By observing the directions given, viz., to have the solution boiling hot, and to use a large excess of H Cl, there is no need of adopting the more troublesome method given in most handbooks, viz., to run in excess of Sn Cl_2, and then titrate back with iodine solution and starch-paste (2 Sn Cl_2 + 2 I = Sn Cl_4 + Sn I_2). By the protochloride of tin method we obtain accurately the amount of Fe_2O_3 present in the ore. Having thus previously determined the total iron, we can obtain, by difference, the amount of iron present as Fe O. In magnetic iron ores the proportion between iron as Fe_2O_3 and iron as Fe O is about 2 : 1. When dissolving ores for iron determinations it is suitable to use an assay-flask with a narrow mouth, into which a rubber-stopper is inserted (p. 47). No oxidation of the Fe O present need then be feared.

c. **Determination of Phosphorus.** Five grams of the finely ground ore are dissolved on the hot iron plate in a beaker in strong H Cl, with the addition of a little H NO_3. After a short heating at a moderate heat redissolve the dry mass in strong H Cl, concentrate by evaporation, add water and filter off the insoluble residue. Then proceed exactly as for a phosphorus determination in steel. If the residue should happen to be very large or look very dark, it may still be suspected to contain some phosphorus. In this case fuse the residue in a platinum crucible with sodium-carbonate, dissolve the mass in H Cl and water, separate the silica by evaporation as usual, test the filtrate for phosphorus with molybdic acid solution, and, if phosphorus be found, add the filtrate to the one first obtained.

d. **Determination of Sulphur.** Five grams of ore are boiled with *aqua regia* and evaporated to gentle dryness on the iron plate. The residue may contain, besides earthy matter and silica, insoluble sulphates, such as sulphates of lime, lead and baryta. The sulphate of lime passes into solution on continued boiling with plenty of water, whilst the other sulphates remain with the residue, which should be tested qualitatively afterwards. After filtering off the residue the sulphur is precipitated and determined exactly as in iron and steel. The residue, if lead and barium be present, should be fused with carbonate of soda. Sodium sulphate, sodium silicate, etc., are thus formed; they can be dissolved with water and the filtrate examined for H_2SO_4 after separating the SiO_2 in the usual way.

e. **Determination of Manganese.** Sometimes, in the case of so-called manganiferous iron ores, a special and quick method for manganese is required. We can here apply the same methods as for steel, using nitric acid and chlorate of potash, or we may proceed according to the directions on p. 51 for high manganese alloys, separating the iron by basic acetates. The only impurities which may interfere seriously with the accuracy of such manganese determinations in ores are baryta and lime. In the nitric acid and chlorate of potash method, with titration, the danger from these causes is comparatively small. In the basic acetate and bromine and ammonia method, however, baryta, if present, must be removed before precipitating the manganese. The baryta is removed by evaporating the H Cl solution of the ore with $H_2 SO_4$, diluting with water and filtering off the sulphate of baryta. The manganese precipitate from the bromine and ammonia process may be freed from contaminating lime by simply redissolving and reprecipitating with bromine and ammonia.

Manganese ores, consisting of the higher oxides of manganese, are much used in the manufacture of the higher alloys of iron and manganese. Such ores may be analyzed according to the methods already given.

f. **Determination of Moisture and Loss on Ignition.** The moisture is determined by drying a large quantity, say 100 grams, of ore at 120° C. and weighing.

The loss on ignition is determined by igniting about one gram of ore in a platinum crucible. The loss represents chiefly water, carbonic acid, organic matter, and possibly some sulphur.

g. **Complete Analysis of Iron Ores.** Half a gram of finely ground ore is fused in a platinum crucible with about 3 grams of carbonate of soda. If the ore be very rich in iron, it is better to fuse the insoluble residue, obtained by treating 5 grams of ore with H Cl, so as to obtain a larger quantity of material in which to determine the small amounts of other elements than iron present. After fusion the mass is dissolved in H Cl and water, putting the crucible into the beaker, to remove all traces of the fusion. When much manganese is present, which is indicated by a deep blue-green color of the mass, it is best to remove the mass and dissolve it in a separate beaker, treating the crucible with acid in another beaker. Hydrochloric acid evolves chlorine with the higher oxides of manganese, whereby the platinum would be seriously attacked.

In the fusion with carbonate of soda, sodium silicate, sodium phosphate, sodium sulphate, sodium aluminate, sodium manganate, etc., are formed; whilst the bases, viz., the iron, magnesium, calcium, some manganese and some aluminium, are separated in the fusing, fluid mass as oxides. By extracting with water only it would therefore be impossible to effect a complete separation of the silica from the bases. The alumina and the manganese will in part go into solution, and in part remain with the residue. We therefore proceed as usual, using H Cl with the water, and separating the silica by evaporating to gentle dryness. The dry mass is then moistened with a little strong H Cl and dissolved in water; the silica is then filtered off and washed with H Cl and water, ignited and weighed.

If the fusion was made on an insoluble residue, as mentioned above, the filtrate from the silica is joined with the H Cl solution previously obtained. In either case the operations are essentially the same.

In the filtrate from the silica make a basic acetate precipitation (p. 31). The filtrate from this contains the manganese, the lime, and the magnesia, whilst the basic acetate precipitate contains the iron, the alumina, the phosphoric acid, the titanic acid, and possibly traces of silica. Copper and lead are only in part carried down with the basic acetate precipitate, whilst antimony and arsenic, if present, accompany the same completely. The basic precipitate is redissolved in H Cl, and again precipitated with ammonia in small excess at a boiling heat. This precipitate is ignited and weighed, then redissolved in H Cl, whereby any titanic acid and silica are left behind. If the iron and phosphorus have been previously determined, we can now obtain the alumina from the difference between the total weight of the precipitate and the $Fe_2 O_3$, the $P_2 O_5$, and the $Ti O_2 + Si O_2$ in the insoluble residue. It is, however, more accurate to make a special determination of the iron in the solution obtained after the separation of the Ti and Si, and for this purpose we make use of a very sharp method, as follows: Nearly neutralize the iron solution last referred to with carbonate of soda, only using such a proportion of same as will contain not more than .25 gram of iron. Add about 4 grams of iodide of potassium, using a small flask with tightly fitting ground-stopper, so that the air can be well excluded. The following reaction takes place: —

$$Fe_2Cl_6 + 2\,KI = 2\,FeCl_2 + 2\,KCl + 2\,I.$$

The liberated iodine dissolves in the excess of potassium iodide. The reaction is promoted by slight warming. We then introduce into the dark-red solution from a porcelain crucible a weighed quantity of pure mercury, say about 6 grams, and shake until the solution has become colorless. The following reaction has then taken place: $Hg + 2\,I = Hg\,I_2$. During the shaking a current of CO_2 should be passed through the solution, to exclude the air; if, however, the stopper fits well, and the solution reaches nearly to the top, the above precaution becomes less necessary. When the brownish color, due to the free iodine, has turned light yellow, a little starch-paste is added, which gives the solution a blue color; and when this finally disappears it is a sharp indication that all the free iodine has combined with mercury. From the above formulas we find that one part of mercury, dissolved as iodide, corresponds to .56 part of iron. The remaining mercury is therefore poured back into its porcelain crucible, dried by contact with filter paper and weighed, and the iron calculated from the loss which the mercury has sustained. Another very good method of determining the liberated iodine is to use a standard solution of thiosulphate of soda, when the following reaction takes place : —

$$2\ Na_2\ S_2\ O_3 + 2\ I = 2\ Na\ I + Na_2\ S_4\ O_6.$$

The thiosulphate must be restandardized so often, however, that its use becomes inconvenient in practice. The filtrate from the basic acetate precipitate is rendered ammoniacal, the manganese precipitated with bromine and determined as usual. The filtrate from the manganese contains more ammoniacal salts than can be conveniently handled. The best manner of removing them is to evaporate in a tall beaker as far as possible without spitting, and then to cautiously add strong nitric acid. Red nitric oxide fumes are given off, and on continued addition of acid and evaporation the ammoniacal salts are rapidly and completely removed. The residue in the beaker is dissolved in a little H Cl and some cold water and rendered slightly ammoniacal. A solution of ammonium oxalate — about 20 cubic centimetres of the concentrated solution — is then added, together with a little more ammonia. Calcium oxalate is precipitated, and magnesium oxalate remains in solution. After settling, which, if the amount of precipitate be very small, requires twelve hours, the calcium oxalate is filtered off and washed, first with cold, then with hot water. The latter should be taken up in a separate beaker, to prevent the precipitation of difficultly soluble magnesium oxalate. A double filter should be used for filtering the calcium oxalate. The calcium oxalate is converted into Ca O by ignition in a platinum crucible and weighed as such. To the filtrate from the calcium oxalate, which must be cold, add some microcosmic salt solution, $\left[\begin{matrix}NH_4\\Na\\H\end{matrix}\right\}PO_4\right]$, with brisk stirring. The microcosmic salt precipitates the magnesia almost instantly as $\left(\begin{matrix}Mg\\NH_4\end{matrix}\right\}PO_4 + 6\,H_2O)$, which settles rapidly to the bottom; filter, ignite, and weigh it as in the determination of phosphorus. The $Mg_2 P_2 O_7$ contains 36.04 per cent of magnesia (Mg O).

We have thus briefly described the most important points in the analysis of iron ores in practice. As is well known, the absolute separation of the elements in minerals is by no means an easy matter, although in practical working the methods above described will in most cases give fairly accurate results.

Amongst the vast number of raw materials of different character which are used in the manufacture of iron, it is only natural that we should occasionally meet with substances that are especially difficult to determine in the ordinary course of analysis. Such, for instance, is the combined occurrence of much titanium and phosphorus in some ores. No thoroughly satisfactory method for the separation of titanium and phosphorus is yet known. The following method will be found useful in most cases. Fuse .5 to 1 gram of ore with sodium carbonate as usual. Dissolve the mass in hydrochloric acid and water and, without separating the silica, make a basic acetate precipitation. The filtrate from this basic acetate precipitation may be thrown away. Redissolve the basic acetate precipitate in hydrochloric acid and precipitate with ammonia. This last precipitate, which contains the iron, silica, alumina, phosphoric and titanic acids, is fused with sodium carbonate. By treating the fused mass with water we effect a separation of the phosphorus and titanium, the latter remaining behind with the ferric oxide as acid sodium titanate, whilst the former passes into solution as sodium phosphate, together with some sodium silicate and aluminate. The residue after treating with water is dissolved in hydrochloric acid and precipitated with ammonia. This precipitate is filtered off, washed and ignited in a platinum crucible, thus rendering the titanic acid insoluble. The iron, alumina, etc. are then extracted with strong hydrochloric acid, leaving the titanic acid, which is again ignited and weighed. The titanic acid should be tested for SiO_2 with hydrofluoric acid.

Some iron ores contain potassium and sodium. A special method must be followed for their determination. If thoroughly pure reagents are used the following method is very suitable. Volatilize the silica in a few grams of ore with HFl and HCl. Make a basic acetate precipitation in the solution of the residue, thereby separating iron, alumina, titanium, phosphorus, arsenic, etc. To the filtrate from the basic acetate precipitation add ammonium sulphide, thereby separating manganese, as well as any zinc, cobalt, nickel and copper, etc., present in the solution. The filtrate from these sulphides is acidified with H Cl, boiled to expel H_2S, and the separated sulphur filtered off. The filtrate from the sulphur is concentrated by boiling, and the ammonia destroyed by evaporating to dryness with nitric acid as described heretofore. In the solution of the residue the lime and magnesia are separated as already shown, taking care to use ammonium phosphate instead of microcosmic salt for the precipitation of the magnesia. The final filtrate from the magnesium phosphate contains the sodium and potassium, as well as phosphoric acid from the ammonium phosphate used. This phosphoric acid is removed by adding a little ferric chloride solution and making a basic acetate precipitation. The filtrate from this contains the sodium and potassium free from fixed impurities. On ignition of the evaporated solution we thus obtain the sum of their chlorides, and on evaporating and igniting the latter with H_2SO_4 we obtain the sum of their sulphates. From these data the amounts of sodium and potassium are easily calculated. Let x be the amount of sodium, y the amount of potassium sought.

$$\text{Sum of chlorides} = A = x + \frac{Cl}{Na}x + y + \frac{Cl}{K}y.$$

$$\text{Sum of sulphates} = B = x + \frac{SO_4}{2\,Na}x + y + \frac{SO_4}{2\,K}y.$$

By introducing the combining (atomic) weights of potassium and sodium respectively we obtain,—

$$A = 2.54\,x + 1.90\,y.$$
$$B = 3.08\,x + 2.23'y.$$

h. **The Determination of some Elements of either rare occurrence, or which occur in very small quantities in Iron Ores, such as Arsenic, Antimony, Lead, Cobalt, Nickel, Zinc, and Barium.** The ordinary qualitative tests for these elements need not be described here; only a few directions for their determination will be given.

Arsenic, Antimony, and Lead. Ten grams of ore are dissolved in HCl and HNO_3, evaporated to dryness and redissolved with HCl and a large quantity of water. The water must be added rapidly, to prevent the separation of basic salts of antimony. The iron is reduced to the state of FeO by means of ammonium sulphite, and excess of sulphurous acid boiled off. Sulphuretted hydrogen is then passed through at a temperature of about 70° C., and the whole allowed to stand for twelve hours in a closed flask. The insoluble substances are then filtered off and the sulphides of arsenic and antimony extracted with ammonium sulphide. If copper be present it is best to use sodium sulphide, as the CuS is not quite insoluble in the ammonium sulphide. The solution of arsenic and antimony is oxidized with $KClO_3$ and HCl, then concentrated and mixed with tartaric acid and ammonia, and finally with magnesia mixture. This precipitates arseniate of magnesia, the antimony being kept in solution by the tartaric acid. In the filtrate from the arseniate of magnesia the antimony is precipitated with H_2S after acidifying, filtered off and ignited in a porcelain crucible with a little nitric acid, yielding SbO_2, containing 79.2 per cent of Sb. The sulphide of lead remains in the residue after treating with ammonium sulphide. It is dissolved by boiling with nitric acid and filtering off from the residue. The filtrate is then evaporated with H_2SO_4 and alcohol added, when sulphate of lead separates.

Zinc, Cobalt, and Nickel. A basic acetate precipitation leaves the above-named three metals in the filtrate. The method for their separation thus becomes very much simplified. Pass H_2S through the filtrate from the basic acetate precipitation, keeping it quite warm. If very *little* free acetic acid be present, some of the Co and Ni are then precipitated along with the Zn S, which, as is well known, is precipitated by H_2S even in excess of acetic acid. By treating the precipitated sulphides with warm acetic acid *all* the Co S and Ni S may be extracted. The Zn S is dissolved in H Cl and precipitated with sodium carbonate in small excess at a boiling heat. The ignited precipitate from this is Zn O, containing 80.26 per cent of zinc. The cobalt and nickel are precipitated, together with manganese, with ammonium sulphide. The Co S and Ni S are freed from Mn S by treating with cold dilute H Cl (1 H Cl, 1.12 specific gravity, to 6 H_2O) which dissolves out the Mn S. The Co S and Ni S are then ignited together with a little nitric acid and ammonium carbonate. The cobalt is hereby converted into metallic cobalt, and the nickel into protoxide (Ni O). A separation of the two metals is entirely unnecessary, owing to the small amounts present of either metal.

Barium. A few grams of ore are fused with sodium carbonate and the mass extracted with water and filtered. Any barium present in the ore is thus left with the residue as barium carbonate. The residue is dissolved in H Cl, and any small amount of silica present separated by evaporation to dryness in the usual way. In the filtrate from the silica the barium is precipitated with $H_2 SO_4$ and weighed as $Ba SO_4$, containing 65.67 per cent of Ba O.

i. **Notes on the Dry Assay of Iron Ores.** The dry assay of iron ores is but little practised since the comparative perfection of the analytical methods in the wet way, but it is still occasionally useful, and a few directions bearing upon the same may be here in place. The aim of the dry assay may be either to ascertain the maximum amount of iron that can be obtained from an ore, in which case many fluxes that are not used in the blast-furnace may be employed, such as fluor-spar, glass free from lead, etc.; or the object may be to determine the practical working of an ore in the blast-furnace burden, which has been previously calculated from results of analysis. The former is generally the aim of the operation. When using fluor-spar a considerable amount of silicon is taken up by the reduced metal. The reaction by which this occurs is considered to be:

ANALYSIS OF IRON ORES.

and
$$12\ Ca\ Fl_2 + 4\ Si\ O_2 = 4\ Ca\ Si\ Fl_6 + 8\ Ca\ O,$$

$$Ca\ Si\ Fl_6 + n\ Fe + 2\ C + Si\ O_2 = Fe\ n\ Si + Ca\ Fl_2 + Fl_4\ Si + 2\ CO.$$

Fluor-spar is thus apt to cause too favorable results, and must be used with caution. The best way of testing an unknown ore in the dry way is to weigh out say five one-gram samples of the ore into five numbered charcoal-lined crucibles, charging each sample with a different proportion of quartz and carbonate of lime. This is the old well-tried Swedish method. As to the choice of crucibles and furnace, there is no need of entering into minute descriptions of the same, as all dealers in chemists' supplies can furnish apparatus of this kind. The mixtures of ores and fluxes are put into the charcoal-lined crucibles, a little flux being put separately on top, in order to wash down any globules of metal from the sides of the crucibles during fusion. A piece of charcoal is placed on top of each crucible, and the fusion proceeded with in a wind-furnace or, better still, in a "Sefströms" furnace. Whatever kind of furnace be used, the heat should be raised gradually, say in three stages, so that the operation be completed in one and a quarter hours. The first raise of heat may be made after half an hour, then after five minutes another slight raise. After ten minutes more full heat is applied for about twenty minutes. The crucibles are then taken out and their contents, after cooling, emptied on to watch-glasses. The buttons of metallic iron are weighed, and any globules contained in the slag extracted with a magnet and weight added. The slag is also examined; the principal observations to be made are whether it seems well fused and separated from the iron, and whether it has a stony or glassy fracture, etc. (more or less basic or acid). No very positive conclusions, however, can be drawn from the colors of these crucible-slags.

B.—*Analysis of Slags, Limestones, etc.*

Slags may be analyzed according to the directions given for iron ores, with very few modifications. All slags contain silica in varying proportions. Blast-furnace slags contain 25 to 65 per cent, Bessemer and open hearth slags 12 to 55 per cent, puddling and similar slags 5 to 35 per cent of silica. Blast-furnace slags contain as bases chiefly calcium, magnesium and aluminium, whilst the other slags contain mostly iron and manganese. Slags from coke blast-furnaces contain much sulphur as calcium sulphide. Such slags can often be completely decomposed by H Cl, and the sulphur in them then determined by the bromine method exactly as in steel. Phosphorus rarely occurs in blast-furnace slags. So-called basic slags, which are chiefly produced by the Thomas' basic Bessemer process, contain up to 30 per cent of phosphoric acid. The determination of phosphorus in such slags is of importance. Such determinations are, however, rendered somewhat difficult by the large amount of phosphorus present. The following method can be recommended.

Weigh out about one gram of the finely ground slag and fuse with sodium carbonate in a platinum crucible. Separate the silica as described in the case of iron ores, and make a basic acetate separation in the filtrate. The basic acetate precipitate is dissolved in H Cl and boiled down with HNO_3 to remove H Cl and acetic acid. The solution is then diluted to say half a litre. From this volume aliquot portions are measured off and the phosphorus determined with molybdic acid and magnesia mixture as usual.

Limestones and other fluxes are easily analyzed according to the same methods that are used for ores, etc. Some judgment is required in weighing out a suitable quantity of sample, so as not to obtain too bulky precipitates.

C.—*Analysis of Coal and Coke.*

Complete analysis of fuel is seldom required in practice. When, however, such analyses have to be made, the platinum apparatus for carbon determination in iron and steel in combination with the platinum-tube used in gas analysis—*vide* below—may be conveniently used.

In practice the determinations most frequently made in fuels are: moisture, volatile matter, fixed carbon, ash, phosphorus and sulphur.

(a) *Determination of Moisture, Volatile Matter, Fixed Carbon and Ash.* Two grams of powdered sample are dried in a weighed platinum crucible at 120° C. for one hour. The loss of weight gives the moisture.

The crucible is then heated, with the lid on, in the flame of a Bunsen's burner, letting the flame completely surround the crucible. The lid should fit tightly. The loss of weight gives the volatile matter. During this heating some of the escaping hydrocarbons suffer dissociation and deposit carbon as a coating on the sides of the crucible and on the under side of the lid. When no more flames appear around the edges of the lid all the volatile matter may be considered as expelled. The fixed carbon is then burnt off with the aid of a slow current of oxygen, which is thrown upon the heated mass through the stem of a clay tobacco pipe. The loss of weight gives the fixed carbon, the ash being obtained at the same time. The ash may now be fused and analyzed as an iron ore or slag.

(b) *Determination of Phosphorus.* The phosphorus may be determined in the ash by fusing the same with carbonate of soda, as described heretofore.

(c) *Determination of Sulphur.* "Eschkas method." Two grams of the powdered sample are mixed with 3 to 4 grams of a mixture of 2 parts of strongly calcined magnesia and 1 part of carbonate of soda, and heated for about one hour in an open platinum crucible, with frequent stirring with a platinum wire. The light, flocculent magnesia causes a rapid combustion, and the sulphur enters into combination with sodium as sulphate, sulphite, sulphide, etc. When all the carbon is burnt out, the contents of the crucible is thrown into a beaker containing hot water. The magnesia is filtered off and washed with hot water. Some bromine water is now added to the filtrate from the magnesia, in order to convert all the sulphur into H_2SO_4. The filtrate is acidified with H Cl, some Ba Cl_2 added and the bromine boiled off. The barium sulphate is then filtered off and determined as usual.

CHAPTER IV.

NOTES ON GAS ANALYSIS.

BLAST-FURNACE gases are chiefly analyzed with the view of ascertaining the relative quantities of CO_2 and CO, as an index of the good or bad working of the furnace, while the composition of gas-producer gases is usually desired in order to ascertain their heating power. The following is a convenient method for the complete volumetric analysis of gases, as they occur in iron and steel manufacture.

a. **Collection of Gases.** A sample of gas may be collected by means of a rubber bag (supplied for the purpose by chemical dealers), from which the air has been previously exhausted by means of the suction pump. A rubber bag is not suitable when the sample of gas has to be preserved for any length of time. The arrangement shown (Fig. VIII.) is then to be recommended. The gas here passes up through the funnel F and the rubber-tubing with the glass tube K, filled with asbestos for retaining dust. A saturated solution of common salt, Na Cl, should be used to fill the two bottles. Such a solution does not absorb any gas, whilst pure water absorbs carbonic acid in considerable quantity. According to Bunsen, the following gas volumes (at 0° C. and 760 millimetres) are absorbed by one volume of air-free water at 15°, on shaking:—

Oxygen	(O)	. . . .	.030
Nitrogen	(N)	. . . .	.015
Carbonic oxide	(CO)	. . . .	.024
Hydrogen	(H)	. . . .	.019
Marsh gas	(CH_4)	. . . .	.039
Olefiant gas	(C_2H_4)	. . . .	.161
Carbonic acid	(CO_2)	. . . .	1.002
Sulphuretted hydrogen	(H_2S)	. . . .	3.300
Sulphurous acid	(SO_2)	. . . .	45.000
Ammonia	(H_3N)	. . . .	727.000
Hydrochloric acid	(HCl)	. . . .	458.000

With increase of temperature the absorbing power of water is diminished. In the following method use is made of water for measuring the gases, but as there is no shaking, and the whole analysis is completed in little more than one hour, no considerable error can ensue from the absorbing influence of the water. When collecting the gas care should be taken that the sample obtained be a fair average representing the gas to be examined; thus in some cases, where the gas is suspected of being irregularly mixed in the flue, etc., the sample should be taken from a tube full of small holes stretching across the whole flue through which the gas is passing.

b. **The Gases to be Determined** are CO_2, O, C_2H_4, CO, H, and CH_4. The nitrogen is taken by difference.

Besides these gases the gas-mixture may contain small quantities of SO_2, H_2S, etc. The gas-mixture may also contain much steam, when coming hot from the furnaces. This steam may be estimated by passing a quantity of the hot gas through a weighed chloride of calcium tube, and measuring the volume of gas thus passed through. The chloride of calcium tube may be inserted between the asbestos tube K and the water bottle (Fig. VIII). The H_2S and SO_2 cannot be determined over water. In this case mercury tubes must be used. A solution of 3 grams of iodine + 4.5 grams of iodide of potash in 50 cubic centimetres of water can be used for absorbing these gases. The SO_2 is thereby oxidized to H_2SO_4, and the H_2S is converted into 2 HI, with separation of sulphur; the iodine solution hereby loses its brown color.

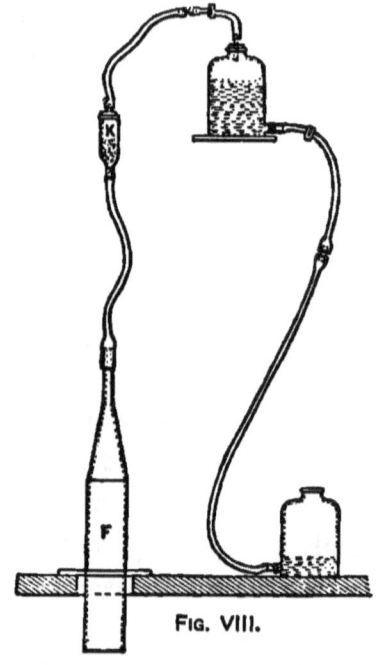

Fig. VIII.

Fig. IX. shows the apparatus for determining the ordinary ingredients in the gases in question. The burettes B and B_1, of 100 cubic centimetres capacity, are graduated into $\frac{1}{10}$ cubic centimetres. B has a funnel at the upper end; between this funnel and B there is a glass stop-cock with two perforations so arranged that B can be connected either with the funnel or with

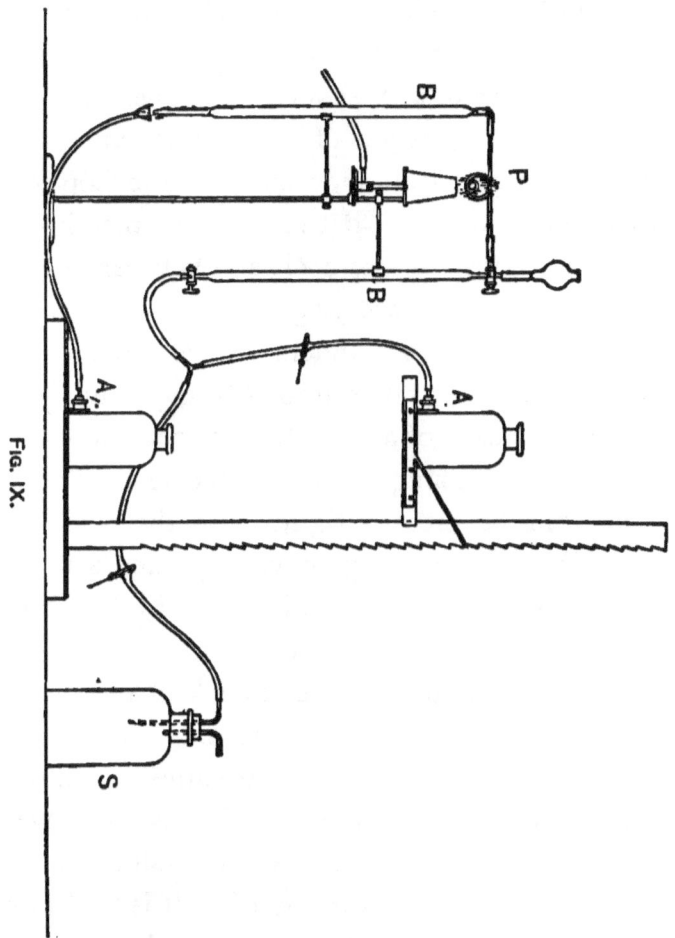

Fig. IX.

B_1 and the platinum tube P.[1] The latter consists of a tube with one-half millimetre internal diameter, twisted

---

[1] J. Bishop, of Malvern Station, Sugartown, Chester County, Penn., supplies all the platinum apparatus recommended in this book.

into a single coil and provided with small cylinders of German-silver at the ends; to these cylinders the rubber-tubings can be securely attached. The lower end of B is connected with the water-bottle A and the suction-bottle S by means of a three-way glass-tube. The air in S is kept rarified by means of the suction-pump, so that liquid can be drawn from B, even when the stop-cock at the upper end is closed. Between B and the three-way tube is a one-way glass stop-cock, and between the three-way tube and A and S are pinch-cocks. The remainder of the apparatus is easily understood from the sketch.

A sample of gas is taken into B through the neck of the funnel by means of a glass tube, to the end of which a piece of rubber-tubing has been attached. The glass tube can thus be pressed tightly into the neck of the funnel. The burette being previously filled with water, the gas is drawn into the same by lowering the flask A. The volume of the gas thus drawn in is then read off by closing the stop-cock at the lower end of B, leaving the funnel connected with B. The ends of B and the funnel should have very narrow apertures. A file-mark is made on the funnel about 60 millimetres above the stop-cock, and the funnel is always kept filled with water up to this file-mark when reading off. It is well to have the gas a little compressed, so that 1 cubic centimetre or so will bubble out through the water in the funnel. The volume of gas should be read off by the lower part of the "meniscus," or water-surface. The temperature is noted, and correction is made for the amount of aqueous vapor corresponding to said temperature as follows. Suppose the temperature be 15°, and the pressure 760 millimetres. From the tension-tables for aqueous vapor (*vide* Thorpe, " Quantitative Analysis," or Bailey, " Chemists' Pocket-book ") we find the corresponding tension to be 12.690 millimetres mercury. Now 1 cubic metre of aqueous vapor, according to Bunsen, at 0° and 760 millimetres weighs .8048 kilos. Thus we have:

$$760 : 12.699 = .8048 : x \therefore x = .0135 \text{ kilos.}$$

By volume this is:

.8048 : 1000 = .0135 : $x \therefore x =$ 16.77 liters at 0°, or at 15° = 16.77 (1 + .004 × 15) = 17.78 liters, or in 100 cubic centimetres of gas (the quantity generally used for analysis) 1.778 cubic centimetres.

At 20° the tension of aqueous vapor is 17.391 millimetres, and its volume in 100 cubic centimetres of gas = 2.40 cubic centimetres. Therefore if 100 cubic centimetres were originally measured off, the actual volume of dry gas would in this case be only 97.6 cubic centimetres. The temperature should be observed in the subsequent operations throughout, as the gas expands according to the formula $v (1 + .004 \, t°)$, or about .4 cubic centimetres for every degree Centigrade.

c. **Determination of Carbonic Acid.** The ingredients of the gases must be determined in the order described, as some of the reagents used for their absorption would otherwise absorb more than one gas.

For the removal of CO_2 is used a solution of 16 grams of potassium hydrate in 100 cubic centimetres of water. Four or five cubic centimetres of this solution absorb 100 cubic centimetres of CO_2. Draw out a few cubic centimetres of water by means of the suction bottle, taking care to have the funnel shut off when the lower stop-cock is open. Let a few cubic centimetres of the potassium hydrate solution flow into the burette from the funnel. All the CO_2 is absorbed when the alkaline liquid flows down slowly along the sides of the burette. The potassium hydrate is then carefully washed out by repeatedly drawing off the liquid from the bottom of B by means of suction, and letting pure water flow in from the top of the burette; the remaining volume of gas is then read off.

d. **Determination of Oxygen.** The absorption of the oxygen is effected by means of pyrogallate of potash. Twenty grams of pyrogallic acid (a light, white powder; must be kept in darkened bottles) are dissolved in 100 cubic centimetres of air-free water. This solution is mixed, immediately before use, with its equal volume of the above K O H solution. Two cubic centimetres of this solution absorb the oxygen in 100 cubic centimetres of air.

e. **Determination of Ethylene ($C_2 H_4$).** Ethylene decomposes at a high heat into carbon and marsh gas (CH_4), and does not generally occur in blast-furnace gases, nor in producer-gases from coke. Producer-gases from bituminous fuel may contain as much as two per cent of $C_2 H_4$.

Fuming sulphuric acid ($H_2 S_2 O_7$), as well as bromine,

absorb the hydrocarbons of the $C_2 H_4$ series; bromine is the more convenient reagent to use. Caution must be observed in this case, as the bromine vapor has a great tension. One hundredth of a gram of bromine absorbs about one cubic centimetre of $C_2 H_4$, forming ethyl-bromide. Water shaken up with bromine contains .02 to .03 grams of bromine per cubic centimetre. A very small amount of bromine water is consequently required for the absorption of ethylene. The ethyl-bromide appears in oily drops on the sides of the burette, when ethylene is present.

f. **Determination of Carbonic Oxide.** The CO is absorbed by means of a solution of 15 grams of $Cu_2 O$ (sub-oxide or red oxide of copper) in 100 cubic centimetres of H Cl, 1.19 specific gravity. The solution is prepared by adding the $Cu_2 O$ to the H Cl at 70°–80° temperature, and allowing a little paraffin to melt on the surface, to prevent oxidization. One cubic centimetre of the fresh sub-chloride of copper solution absorbs about 20 cubic centimetres of CO. It is supposed that the compound 2 Cu Cl + CO is hereby formed. In washing out the sub-chloride of copper some dilute H Cl must first be used, to prevent the separation of the white chloride.

g. **Determination of Hydrogen and Marsh Gas.** After removing the above gases, about 20 cubic centimetres of pure oxygen are taken into B, the gas having been previously transferred to B_1, through P, by means of the bottles A and A_1. The gas is then brought back into B_1, and the total volume read off. Twenty cubic centimetres of oxygen is an ample quantity for the gases in question. As a safeguard against explosions, the rule may be observed, that *the sum of the volumes of the gases taking part in the combustion must not exceed one-half of the total volume of gases in the burette.* The platinum coil is now heated to redness by means of a small Bunsen burner, whilst the gas is passed from B to B_1, and again from B_1 to B. Complete combustion of hydrogen and marsh gas then takes place in the tube. Vapor of water, which condenses, and carbonic acid are formed, one volume of hydrogen giving one volume of water, and one volume of marsh gas giving two volumes of water and one volume of carbonic acid, according to the following formulas:

$$\frac{2\ H}{2\ \text{vol.}} + \frac{O}{1\ \text{vol.}} = \frac{H_2 O}{2\ \text{vol.}}$$

$$\frac{CH_4}{2\ \text{vol.}} + \frac{4\ O}{4\ \text{vol.}} = \frac{CO_2}{2\ \text{vol.}} + \frac{2\ H_2 O}{4\ \text{vol.}}$$

The actual number of cubic centimetres, representing the *free* hydrogen in the original gas volume, is consequently obtained from the formula,

$$\text{Hydrogen} = \tfrac{2}{3}\ [M - 2\ G],$$

M being the *total* diminution of volume after combustion, and G the volume of carbonic acid from the marsh gas. This carbonic acid is determined as previously described, giving the volume of marsh gas direct.

h. **Notes on Products of Combustion** (Eggertz). When fuel is properly burnt the escaping products of combustion should not contain any combustible gases such as hydrocarbons, hydrogen or carbonic oxide, but only carbonic acid, nitrogen, oxygen and steam. To obtain this result a certain excess of air is generally required, but too great an excess must be avoided, as otherwise heat is carried away uselessly. As a rule it is supposed that the excess of air should be equal to the theoretical or calculated amount of air necessary for complete combustion.

The actual excess of air in each separate case can be estimated by determining the amount of carbonic acid present in the products of combustion. Experiments made in Munich* have shown that the following figures give the relation between the excess of air used and the carbonic acid present with the products of combustion:

4 per cent CO_2	4.6	times the theoretical air quantity.
5 " "	3.5	" " " " "
6 " "	3.	" " " " "
7 " "	2.5	" " " " "
8 " "	2.3	" " " " "
9 " "	2.	" " " " "
10 " "	1.7	" " " " "
12 " "	1.5	" " " " "
17 " "	1.0	" " " " "

The carbonic acid occupies the same volume as the oxygen that enters into the same; but oxygen has also

---

* Bayrisches Industrie und Gewerbeblatt, 1880.

been consumed for the combustion of hydrogen, hydrocarbons and sulphur.

The best result in the above experiments was obtained when the gases contained 10 per cent. of CO_2, corresponding to 1.7 times the theoretical amount of air. The loss of heat through the combustion products hereby went down to 10 per cent, whereas with 4 per cent CO_2 the loss thus incurred was 36 per cent, and with 8 per cent CO_2 18 per cent. If the CO_2 exceeds 10 per cent it is to be feared that some combustible gases may yet be present in the escaping gases. In the best cases these contain only 1 per cent. of CO and H, but in bad cases as much as 3 per cent. CO and 1 per cent H, or more.

For the combustion of 1 kg. of pure carbon to carbonic acid are required 2.67 kg. oxygen or 11.56 kg. air. If, as usual, we assume that at least the double theoretical amount be required for complete combustion, 1 kg. of pure carbon will require 23 kg. or 18 cubic metres of air.

For measuring the amount of air rushing into a furnace use is made of anemometers, obtainable from scientific instrument makers.

The loss of heat caused by the products of combustion is due to three circumstances: 1st, incomplete combustion of the fuel, so that combustible gases escape; 2d, the amount and the temperature of the products of combustion; and 3d, the accompanying steam.

To calculate these losses we make use of the table

NOTES ON GAS ANALYSIS.

on p. 87. The specific heats per cubic metre of air, oxygen, hydrogen, nitrogen and carbonic oxide are in these calculations considered identical and expressed by the number, .307.

Example. Chimney gas, temp. = 210° C.

Carbonic acid = 12.5 per cent.
Carbonic oxide = 2.3 "
Hydrogen = 1.0 "
Oxygen = 4.2 "
Nitrogen = 80.0 "

100.0 volumes.

One hundred cubic metres of this gas are accompanied by 8 kg. of water as steam.

The combustible gases consist of CO and H, and these are capable of developing:

2.3 × 3007 + 1.0 × 2655 . . . . = 9,571 h. u.
with 100 c. m. of gas at 210° C. are carried off (2.3 +
1.0 + 4.2 + 80.0) .307 × 210 + 12.5 × .4256 × 210 = 6,758 h. u.
8 kg. water as steam occupy a volume of 8 : .8048 =
9.940 c. m., and this carries off 9.940 × .3823 × 210 = 798 h. u.

Total loss of heat 17,127 h. u.

This heat can be produced by the combustion of $\frac{17127}{8080}$ = 2.12 kg. pure carbon.

In 100 c. m. gas are contained (12.5 + 2.3) × .5363 = 7.94 kg. pure carbon. Thus of the 7.94 kg. pure carbon in the fuel, 2.12 kg., or 26.94 per cent, have been lost.

i. **Calculation of the amount of Air blown into a Blast Furnace** (Stockmann, Beckert).

Example: A blast furnace produces daily 40 tons of pig-iron with 4 per cent carbon, using 1.3 tons coke per ton pig-iron. The coke contains 77 per cent carbon; the ores are free from carbonic acid; the limestone (48 tons) has 43 per cent CO_2 = 11.73 per cent C. The furnace consequently received

with the coke,	$40 \times 1.3 \times .77 =$ 40.040 tons carbon	
with the limestone,	$48 \times .1173 =$ 5.630 " "	
	total 45.670 " "	
The pig-iron contains	$40 \times .04 =$ 1.600 " "	
Consequently are volatilized,	44.070 " "	

or per minute 30.60 kg.

The escaping blast-furnace gas contains:

	Vol. %.	Weight %.
Nitrogen	55.76	54.79
Carbonic acid	9.99	15.42
Carbonic oxide	24.88	24.45
Marsh gas	.40	.22
Hydrogen	.97	.07
Steam	8.00	5.05
	100.00 vol. per cent.	100.00 weight per cent.

This gas contains in the CO_2 4.21 per cent, in the CO 10.48 per cent, in the CH_4 .17 per cent, total 14.86 weight per cent of carbon; consequently the carbon in the gas is to the nitrogen as 14.86 to 54.79. With the 30.6 kg. of volatilized carbon 112.8 kg. of nitrogen consequently leave the furnace per minute, corresponding to 146.85 kg. or 113.57 cubic metres of air.

Another ready way of calculating the air is given by Ledebur. If A be the consumption of fuel per 24 hours, p the amount of carbon per kg. of fuel, and Q the consumption of air per minute, we have

$$Q = \frac{Ap}{320} \text{ cubic metres.}$$

This formula must by multiplied by a coëfficient .75 to .85 to correct for errors. The more difficultly reducible the ores are the smaller need the coëfficient be.

We can also calculate the amount of gas leaving the blast furnace in the same way as the incoming air. In the above example we find that for every kg. of gaseous carbon .705 kg. are present as CO, and that 30.6 kg. of carbon leave the furnace per minute. Out of these 30.6 kg. 21.573 kg. consequently leave the furnace as CO, weight about 50.3 kg. = 40.23 c. m. With 24.88 per cent CO by volume in the gases, we thus readily find that 162 cubic metres of gas leave the furnace per minute.

l. **Calculation of the amount of Carbon used for direct and indirect reduction in the Blast Furnace from the relation between CO_2 and CO ; $\frac{CO_2}{CO}$ by vol.**
$= k ; \frac{CO_2}{CO}$ by weight $= m = 1.57$ k.

Whatever degree of oxidation the iron may possess in the ore when charged on the blast furnace, it will ultimately arrive at the lowest degree, Fe O, before being reduced to metallic iron. The Fe O does not *rapidly* give up its oxygen except at a very high temperature, at the same time absorbing per unit iron the same number of heat units as are set free when the same unit of iron combines with oxygen to form Fe O. At the high temperature at which the reduction of Fe O takes place rapidly the stability of CO_2 is exceedingly small, and even if the reduction at first take place with CO:

$$Fe\,O + CO = Fe + CO_2,$$

the CO_2 formed would take up carbon and form CO, thus: $CO_2 + C = 2\,CO$, which means loss of heat and waste of carbon in the blast furnace, since the heat generated by the oxidation of the carbon is much less than the heat absorbed by the reduction of Fe O. Thus the reduction by carbon direct, either:

$$Fe\,O + C = Fe + CO,$$
or
$$Fe\,O + CO + C = Fe + 2\,CO,$$

must be avoided. In order to accomplish this the stability of the CO_2 generated by the reduction with CO must be insured by injecting a sufficient excess of air. The oxygen in this air forms CO, part of which reduces the Fe O, the remaining part protecting the CO_2 formed from being decomposed. The gas mixture then ascends up through the furnace, meeting the mixture of ore, fluxes and fuel, thereby giving off its heat and reducing the ores, etc.

Professor Akerman has made experiments on the reduction of ores by various gas mixtures, according to which carbonic oxide must not be mixed with more than half its volume of carbonic acid, to obtain a strongly reducing action on Fe O at a temperature of 800°–900°.

Thus the following formula would indicate the minimum amount of carbon requisite for reduction:

$$3\ CO + Fe\ O = 2\ CO + CO_2 + Fe,$$

or per atom iron 3 atoms of carbon (3×12 parts by weight of carbon per 56 p. b. u. of iron). If the gas mixture thus generated meets $Fe_3 O_4$ higher up in the furnace, this reaction would then take place:

$$6\ CO + 3\ CO_2 + Fe_3\ O_4 = 5\ CO + 4\ CO_2 + 3\ Fe\ O,$$

and if this gas mixture meet $Fe_2 O_3$ still higher up in the furnace, the reaction would be:

$$10\ CO + 8\ CO_2 + 3\ Fe_2\ O_3 = 9\ CO + 9\ CO_2 + 2\ Fe_3\ O_4,$$

the gases finally leaving the furnace with equal volumes of CO and CO_2. For reducing one weight of iron the minimum amount of carbon required would thus be .643 weights. In reality many blast furnaces show less consumption of carbon than that, even with more CO in the gases than as per above formulas. This can be explained by the fact that there is a surplus of heat in the blast furnace, partly generated in the furnace and partly introduced with the heated blast. This surplus of heat covers the loss caused by direct reduction, thus effecting a saving of coal.

It should be borne in mind that the chemical reactions which take place in the blast furnace are nearly all more or less interwoven, forming no distinctly defined stages, as would seem from the above formulas.

Thus the CO_2 formed by the reduction of iron oxides will give off oxygen to carbon more or less through the whole furnace. Such is the disadvantage considered to be derived from this fact that it has been proposed to separate the fuel and the ore in the blast furnace, not allowing them to meet until carbonization is to take place.

In calculating the amount of carbon used for direct reduction we must deduct the CO_2 from ores and fluxes. Another source of error is the dissociation of CO, which takes place in the upper part of the furnace: $2\ CO = C + CO_2$.

If the ore used be Fe_3O_4, the relation between $CO_2 : CO = m$ should be 1.25, and if the ore be Fe_2O_3, m should be 1.57, for complete indirect reduction, according to Prof. Åkerman's formulas.

m. **Notes on Heat Calculations.** In the appendices are given tables for facilitating the calculation of the calorific power and theoretical temperature of gases; supposing the gases to be all of 0°C. and 760 m. m. pressure, according to the formula:

$$W = T \text{ (combustion products} \times \text{ their resp. spec. heats)}.$$

If the gas had an initial temperature t, the formula would be

NOTES ON GAS ANALYSIS. 93

$W +$ (ingredients of gas × their resp. spec. heats) $t = T$ (combustion products × their resp. spec. heats).

The theoretical temperatures are of course never attained in practice owing to dissociation and other causes, but they are of value for comparing different gases, etc.

For solid fuel we have the formulas:

$80.8 \times C + 344.6$ H (Scheurer Kastner) and $80.8 \times C + 344.6 \times \left(H - \dfrac{O}{8}\right)$ (Dulong), where C, H and O mean the resp. per cent of carbon, hydrogen and oxygen.

This formula requires a complete elementary analysis.

An easier way of calculating the comparative heating effect of different solid fuels is the Berthiers test. In this test the amount of lead reduced from lead oxide by the different fuels is used for comparison.

j. **Analyses of gases, per 100 vol.**

	C_2H_4	CO_2	O.	CO.	H.	CH_4	N.
Blast furnace gas (charcoal + coke).	—	2 @ 5	—	10 @ 35	.1 @ 8	0 @ 2	55 @ 65
" " (coal)	0 @ 1.6	3 @ 17	—	16 @ 31	.5 @ 12	.1 @ 8	50 @ 66
Producer gas (charcoal + coke)	—	5 @ 7	—	23 @ 35	1.5 @ 4	0 @ 2	63 @ 65
" " (coal)	0 @ 2	4 @ 8	0 @ 3	20 @ 25	2.5 @ 8	1 @ 2.5	60 @ 68
" " (wood and peat)	0 @ 2	5 @ 11	0 @ 1	17 @ 31	6 @ 15	.5 @ 6	51 @ 59
" " (water gas)	—	1 @ 7	0 @ .8	35 @ 40	44 @ 53	3 @ 5	1 @ 9
Products of combustion	—	4 @ 15	.5 @ 16	0 @ 5	0 @ 3	—	79 @ 81
Bessemer converter gas	—	.1 @ 9	0 @ 8	0 @ 29	0 @ 2	—	65 @ 90
Illuminating gas	4 @ 13	.1 @ 4	0 @ .7	4.5 @ 9	21 @ 55	33–50	.5 @ 5
Products of distillation of pine wood.	1.8	31.8	1.6	34.4	7.1	16.8	6.5
" " " " peat	1.4	41.3	1.3	16.1	19.6	17.1	3.2
" " " " can'l coal	13.6	2.7	1.9	3.6	13.9	58.0	6.3
Natural gas, mostly CH_4, with some CO_2 C_2H_4, O, H, N, etc.							
Air gas, or gas obtained by blowing air through gasoline	1	—	14	—	2	27	56

k. Heating Properties of various Gases (Fred. Taylor).

Nature of the Gas.	Where sample was taken from.	Chemical Composition.	Theoretical flame temperature (Centigrade) of the gas when burned with just enough air to insure perfect combustion.	Theoretical flame temperature (Centigrade) of the gas when burned with pure oxygen.	Heat units (Kilogramme Centigrade) developed per cubic metre of gas consumed.	Number of heat units required to raise the temperature of the gases, resulting from the complete combustion of one cubic metre of the combustible one degree in temperature.		Relative number of cubic metre of the various gases which have to be used to produce the same quality of heat (not degree of sensible temperature), taking natural gas as a standard.
						Burned in Air.	Burned in Oxygen.	
Natural Gas of Pittsburgh.	Analysis made by Dr. G. Hay.	Co = 1. vol. C.H$_4$ = 95. " H = 2. " O = 1.30 "	2,333° C.	Flame temperature of Marsh gas which is about the same as this gas=7,133° C.	8141.	3.489	1.289 about	1.
Water Gas.	Sample of gas taken from Low's gas producers after passing through purifier at the Novelties Exhibition, 10/16/85.	Co = 44.5 vol. O = 50.9 " O = .7 } air. N = 2.8 } 1.1 undetermined.	2,764° C.	6,849° C.	2688.50	.9736	.3925	3.02
Siemens open Hearth Gas.	About an average sample taken from Siemens producers at Midvale Steel Co., 1883.	Co$_2$ = 1.5 vol. Co = 23.6 " H = 6.0 " CH$_4$ = 3.0 " N = 65.9 "	1,600° C. to 1,800° C. according to the quality of the gas.	2,700° C. to 3,000° C. according to the quality of the gas.	1123.41	.6227	.1673	7.25
Marsh Gas.		CH$_4$	2,331° C.	7,133° C.	8482.	3.645	1.189	.96
Carbonic Oxide		Co	2,894° C.	7,075° C.	3207.	1.039	.425	2.7
Hydrogen Gas.		H	2,665° C.	6,944° C.	2665.	.996	.3823	3.05

CHAPTER V.

METALLURGICAL NOTES AND PRACTICAL USES OF THE RESULTS OF ANALYSES.

The Chemical Analysis is of great value not only technically, but also commercially, as regards pig-iron. The analysis determines to a great extent the suitability of ores, fluxes, etc., for certain kinds of pig-iron and other alloys of iron made in the blast furnace, and it determines the suitability of various kinds of pig-iron for different purposes, such as castings, Bessemer, Siemens-Martin, Thomas, crucible, puddling and other processes.

The chemical analysis is of great use for determining, or at least for forming an idea of, the refractory properties of sands, clays, etc.

We have already described the uses of gas analysis for various purposes, and in the appendices are described several practical applications of gas as well as other analyses.

For wrought-iron and steel the chemical analysis cannot so reliably predict the physical properties as for pig-iron. This has already been pointed out in the first chapter of this book. But it *may* do so in many cases, and it is otherwise exceedingly useful in explaining the causes and tracing the sources of defects in steel. The method given in Chapter II. for the determination of "slag and oxide of iron" has found a special application for distinguishing wrought-iron from steel where doubt as to this has existed, the wrought-iron containing much more slag and oxide of iron than the steel.

a. **Pig-iron.** The suitability of an ore for a certain kind of pig depends upon several circumstances besides the chemical composition, such as physical properties, supply of suitable ores of other kinds, etc.

In selecting ores it should be remembered that all the phosphorus goes into the pig. Only with a very hot furnace and a basic burden can any phosphoric acid be brought into the slag. The sulphur, on the other hand, can be removed to a great extent under similar circumstances, passing into the slag as Ca S. The manganese has a very great tendency to pass into the slag, its affinity for oxygen being so great, that ferromanganese can only be produced at temperatures higher than the temperature at which the metal manganese is volatilized. Spiegeleisen and ferromanganese never contain any appreciable amounts of sulphur, the sulphur forming Mn S with the Mn; the Mn S is not soluble in the molten alloy and passes into the slag. Silicon is readily reduced by carbon at high temperatures in the presence of metallic iron. The pig-iron becomes more carboniferous the more manganese and the less silicon and sulphur it contains; even phosphorus acts against high carbon, inasmuch as it makes the metal more fusible, enabling it to melt without taking up much carbon. Rapid driving of a blast-furnace counteracts high carbon.

As to the selection of ores, the *proportion* between the iron and the phosphorus in the ore is of great importance. Carbonates, hydrates and ores containing much sulphur are generally roasted before using; magnetic iron-ores are also generally roasted in order to convert Fe_3O_4 to Fe_2O_3, which is easier to reduce. In the appendices some further remarks on the selection of ores will be found.

The amount of pig-iron from a blast furnace obtained is often calculated according to the formula:

$$\text{Pig} = Fe + \frac{Mn}{2} + \frac{Fe + \frac{Mn}{2}}{20}$$

which means that the pig contains all the iron, half of the manganese and metalloids to the extent of 5 per cent of the metals. For very phosphoriferous pigs the following formula comes nearer the truth:

$$\text{Pig} = Fe + \frac{Mn}{3} + \frac{Fe + \frac{Mn}{3} + P}{30}$$

For castings the selection of suitable pig-irons is of great importance. Gray pig-iron is generally used for castings, white or mixed pig being used chiefly for chilled and malleable castings. The user of cast-iron must know the physical properties of the material as well as the behavior of the material in melting. The latter can be predicted by means of the chemical analysis of the pig. The manganese, silicon and carbon are more or less oxidized in melting, whilst the iron in the presence of those easily oxidized elements remains in the metallic state. The relative avidity with which the three elements unite with oxygen is as in the above order. The temperature influences the oxidation of carbon; at very high temperatures carbon takes up oxygen even at the expense of manganese and silicon. The respective amounts of the elements have here, as in all chemical reactions, much to do with the result. The phosphorus remains with the iron, and it is thus easily understood that its per cent should become somewhat raised. (Phosphorus can be removed from pig-iron according to a process devised by Bell and Knapp, by melting it at a low temperature together with pure iron-ores. The pig for this process ought to contain some manganese, which helps to protect the carbon and silicon from oxidation:) The manganese protects the silicon during melting from oxidation, and should therefore always be present to some extent in order to prevent the pig from losing its gray texture. On an average the manganese in good pig-iron for castings does not exceed 1 per cent, the silicon 2–3 per cent, the phosphorus .5 per cent, the graphite and combined carbon together about 4 per cent. The respective amounts of graphite and combined carbon have much

influence on the physical properties of pig-iron.*
Phosphorus renders the pig not only more easily fusible but also more fluid; manganese has the opposite effect.

For the Bessemer process the pig-iron must not contain too much phosphorus, as the later is not removed in the " acid " process. The Bessemer process is carried out rather differently in different countries; one generally mentions the English, the German and the Swedish process. The chief characteristics of these three processes are for the English process, a not very superheated pig-iron, rich in carbon and silicon but with little manganese; yields a metal low in silicon. For the German process, a superheated pig-iron with much manganese and silicon; yields a metal with low carbon but much silicon (*vide* Chapter I. under " Influence Phosphorus.") For the Swedish process, a hot pig from the blast furnace with much manganese and carbon; but not a *very* high per cent of silicon, and remarkably low in phosphorus; yields a metal low in both manganese and silicon. The initial temperature of the pig when charged in the converter has very much to do with the final composition of the product (*vide* above). Finally, perhaps, the Clapp-Griffith's process should be mentioned, as a modification of the Bessemer process. In the Clapp-Griffith modification a soft steel is principally made, rich in phosphorus, but low in both carbon and silicon. Very few analyses have been published of

---

* Dudley's (Chas. B.) *Transaction Min. Engineers.*

metal made by this process, but it seems to be of good quality, although this is probably due to the absence of carbon, silicon acting indifferently toward phosphorus, according to the German experience.

To demonstrate the influence of the composition of Bessemer pig-iron on the process, some data are here given according to Professor Ledebur, Favre and Silbermann, a. o. 1 kg. Fe burnt to Fe O develops 1352 h. u. giving 1.28 kg. Fe O and .94 kg. nitrogen. Specific heat of Fe O = .20, of N = .25. Initial temperature of the iron = t. Consequently the molten iron possesses an initial heat of .18t h. u., if .18 be the specific heat of iron between fig. 0° and t°. By the products of combustion are taken up (1.28 × .20 + ·94 × .25) t h. u., and the total number of the heat units that benefits the iron-bath is therefore = 1352 + .18t − .491t = 1353 − .311t for every kg. of iron burnt to Fe O. If t be = 1500° we thus find W = 886 h. u., and if we put the specific heat of iron at 1500° = .20 we find the increase of temperature caused by the combustion of 1 per cent of iron = $\dfrac{8.86}{.20}$ = 44 per cent only.

1 kg. Mn to 1.29 kg. Mn O + .97 kg. N develops about 2000 h. u.
" C " 2.33 " CO + 4.47 " " " 2473 "
" Si " 2.14 " Si O₂ + 3.82 " " " 7830 "
" P " 2.29 " P₂O₅ + 4.00 " " " 5760 "

      Specific heat of Mn = .18, of Mn O = .20
      " " " C = .25 " CO = .25
      " " " Si = .18 " Si O₂ = .19
      " " " P = .18 " P₂O₅ = .25

By similar calculations as those for iron we find,

for Mn, $W = 2000 - .42t.$
" C, $W = 2473 - 1.45t.$
" Si, $W = 7830 - 1.18t.$
" P, $W = 5760 - 1.39t.$

or if $t = 1500°$, for the combustion of 1 per cent of either of the above elements, a raise of temperature,

for Mn, of 69° C.
" C, " 6° C.
" Si, " 300° C.
" P, " 183° C.

It is evident, from the above figures, that the carbon cannot furnish enough heat to carry through the Bessemer process, silicon and, in the basic process, silicon and phosphorus being the principal heaters. The facility with which good soft steel is made by the basic process depends largely upon the heating properties of phosphorus, which burns in the last stage of the process when the metal has become more refractory and needs much heat. Pig-iron for the basic process contains generally from 2 — 3 per cent of P, .5 to 2 per cent of Mn and 1.19 — .5 per cent of Si.

Pig-iron for the Siemens-Martin process may be either gray or white, but should have much carbon. The most suitable pig-iron for the open-hearth process is generally a pig containing little P and S, but 2 — 5.5 per cent Si, 3 — 3.5 per cent Mn and 3.5 — 4 per cent C.

Pig-iron for the carbonization in crucibles should contain nearly only carbon and iron; such pig is obtainable from some famous furnaces in Sweden.

b. **Steel.** It has already been mentioned that chemical analysis alone cannot predict the suitability of a steel for different purposes.

But the analysis has done much to remove existing mystery and prejudice concerning the crucible and other processes.

A characteristic property of crucible steel is its large percentage of silicon, which varies between .3 and .5 per cent, and seldom goes down to .1 per cent. The manganese is seldom over .3 per cent; the carbon varies according to the different purposes for which it is intended. The high silicon is by no means beneficial to the quality of crucible steel, and crucibles containing more alumina and less silica would seem to be the remedy for this evil. It is claimed for the crucible steel that it contains less gases than steel made by other processes.

The Bessemer product varies in composition according to the different purposes for which it is intended, and also according to local circumstances. A few examples are here given:

			C	Si	Mn	P
Hard steel	from	Gras, Austria:	1.03	.02	.25	.09
Tool steel	"	Fagersta, Sweden:	.70	.03	.26	.025
" "	"	Midvale, U. S. A.:	1.00	.10	.27	.027
Axle "	"	Leraing, Belgium:	.49	.09	.60	.07
Rail "	"	Osnabrück, Germany:	.19	.50	.87	.14
" "	"	Bethlehem, Pa.:	.35	.05	.75	.08
Steel for plates "		Fagersta:	.09	.01	tr	.025
Thomas steel	"	Vitkowitz, Austria:	.10	tr	.20	tr
Soft steel for nails,		Hofors, Sweden:	.10–.15	.02	.14	.027

The Siemens-Martin metal also varies according to purpose.

The so called "Mitis" process, which has lately come into use, gives very good castings of the softest iron steel. According to information received, the castings are obtained solid by heating the soft scrap used as raw material to the point when it just melts, and then adding an aluminium alloy, which lowers the melting point and causes the metal to become super-heated. The melted scrap does not take up any gases, being heated only to the melting point before adding the aluminium. Aluminium has thus become of interest in the chemistry of iron, and it can readily be determined in steel by a method similar in part to the one given in Chapter II. for titanium.* It may also be determined by difference after estimating the amounts of other ingredients.

As to the influence of the various elements on the properties of steel, this has already been alluded to in the first chapter. We may add that the effect of manganese is about $\frac{1}{6}$ that of carbon. Manganese seems to increase the facility of iron to absorb and dissolve gases. Phosphorus, silicon and carbon protect iron and steel against rust, whilst manganese and sulphur promote rust. Steel has more tendency to become rusty than wrought-iron, which is probably due to the larger amounts of manganese and sulphur in the former.

The changes of steel when heated and cooled are

---

* It should be borne in mind that an equal amount of iron must be present to effect a rapid and complete precipitation of the aluminium by basic acetate.

many and important. Mr. Brinell, of Fagersta, has made experiments* on the changes of texture, and has come to the following conclusions :

1st. When steel loses its coarse crystalline texture without mechanical treatment, this is always accompanied by the transformation of carbon from cement to hardening† carbon or vice versa. The change of texture is exclusively due to the change of state of carbon.

2d. The coarse crystalline texture disappears completely only when the carbon *during heating* changes from cement to hardening carbon. In accordance with this the most coarsely cyrstalline, hardened or unhardened, steel loses this texture completely if heated just to the point where the cement carbon changes to hardening carbon.

3d. In order to convert the carbon into cement carbon in white-hot steel, it must be cooled to a point lower than the point to which unhardened steel must be heated in order to have the carbon converted into hardening carbon.

4th. The change to hardening carbon takes place rapidly at the proper heat.

The change to cement carbon takes place slowly either during the heating up or during the cooling.

---

* The steel for trial had $C = .52$, $Si = .13$, $P = .026$, $Mn = .48$.

† Steel containing only hardening carbon, when treated with dilute nitric acid, gives a sootlike carbon, brown on paper; cement carbon a bluish glistening carbon, black on paper.

5th. *Heat is always set free* when hardening carbon changes into cement carbon, and probably, therefore, heat is absorbed when the opposite reaction takes place.

6th. When the hardening carbon, either during cooling or heating, has been completely changed into cement carbon the texture suddenly becomes coarsely crystalline, and more so, the more coarsely crystalline it was before.

7th. *Rapid cooling can never produce a fine texture in a previously coarsely crystalline steel.* It only fixes the texture existing before the cooling.

8th. The change from hardening to cement carbon requires a suitable heat *and time,* whilst the reverse reaction seems to depend exclusively upon the degree of heat. This is the reason why a rapid cooling prevents hardening carbon from changing into cement carbon.

9th. For the crystallization of steel time is required. If the steel be rapidly cooled the development of crystals is thus checked.

At a "blue heat" all kinds of steel seem to be very coarsely crystalline and brittle. Mr. Brinell's experiments confirm this. When steel is gradually heated its color changes about as follows with the temperature:

Yellow,	220° C.	purple red,	275°
Dark yellow,	240° "	violet,	285°
Brown yellow,	250° "	bluish,	293°
" red,	265° "	light blue,	315°
		gray,	330°

To illustrate the changes of strength of steel with the

carbon, the other composition being the same, the following figures are given :

Lbs. p. sq. in.
Carbon = .55, tensile strength 106,000, elong. per cent. 32 per cent.
" = .65, " 117,000, " " 24 "
" = .75, " 126,000, " " 19 "
" = .85, " 140,000, " " 13 "
" = .95, " 140,000, " " 13 "

c. **Notes on refractory materials.** Silica, Alumina, lime and magnesia, iron peroxide and tetroxide and clay, as well as graphite, are the principal refractory materials used in iron-steel making. The clays are very variable in chemical composition. They consist of aluminium hydrate and silicate, with varying amounts of alkalis, lime, magnesia, iron hydrates, etc. When the sum of the latter ingredients amounts to more than ten per cent., the clay cannot longer be considered refractory. The silica occurs not only in combination with alumina, but also in the free state. As an illustration two extremes are here given :

	Clay 1.	Clay 2.
Al_2O_3	36.30	28.05
Comb. SiO_2	38.94	30.71
Free SiO_2	4.90	27.61
Foreign ingred.	1.26	4.75
Loss on ignition	17.78	8.66

CHAPTER VI.

Notes on Electrolysis.* The electrolysis is beginning to find application not only to fluids but also to melted minerals.

The most striking advantage of electrolysis is the pure state in which metals can be separated by means of the same. It is beyond the province of this book to discuss the practical applications of electrolysis, and only some important principles upon which its application is based can be communicated.

Berthelot has given the following laws: 1st. The heat developed in a chemical reaction is a measure of the physical or chemical work performed—*molecular work*.

2d. If a given combination of simple or compound bodies undergoes a change without external mechanical force the heat developed or absorbed depends *only* upon the initial and final conditions of the combination, of whatever kind the intermediate changes may have been —*equivalence between heat and chemical change*.

3d. Every *chemical* change which takes place spontaneously results in the production of such compounds in the formation of which the most heat is developed.

Heat is developed when elements combine chemically, and if this heat be known, the amount of force necessary

---
* Balling.

for decomposition can be calculated. Favre and Silbermann found the following number of heat units in the formation of the compounds enumerated:

					For 1 Equiva.
For 1 part of iron to			Fe O	1353	75656
"	"	"	Fe_2Cl_6	1745	196170
"	"	"	$FeCl_2$	1775	99302
"	"	"	Fe S	634	35506
"	"	Zinc	Zn O	1291	83915
"	"	"	$ZnCl_2$	1529	101316
"	"	Copper	Cu O	684	43770
"	"	"	$CuCl_2$	961	60988
"	"	"	Cu·S	285	18266
"	"	Lead	Pb O	266	55350
"	"	"	$PbCl_2$	430	89460
"	"	"	Pb S	92	19112
"	"	Tin	SnO_2	1147	135360
"	"	"	$SnCl_4$	1079	126888
"	"	Silver	Ag_2O	57	12226
"	"	"	Ag Cl	322	34800
"	"	"	Ag_2S	51	11048

According to Thompson, the equivalent numbers of heat units corresponding to the formation or decomposition of various compounds is as follows, in aqueous solutions:

$AuCl_3$	27270	Ag_2SO_4	20390
$MnCl_2$	128000	$AgNO_3$	16780
$ZnCl_2$	121250	$CuSO_4$	55960
$ZnSO_4$	106090	$CuCl_2$	62710
$FeCl_2$	99950	$Cu\,2NO_3$	52410
Fe_2Cl_6	255420	$Pb\,2NO_3$	69970
$FeSO_4$	93200	$Pb\,2C_2H_3O$	65760
$Fe_2\,3SO_4$	224800		

NOTES ON ELECTROLYSIS.

The metals develop, according to Thompson, the following numbers of heat units when entering into the following solid compounds:

Cu to	$Cu_2 O$:	40810.₂	$Cu_2 S$:	18069.	Cu Cl :	65750
"	Cu O :	37160.			$Cu Cl_2$:	51630
Fe	Fe O :	99232.	Fe S :	35504.	$Fe Cl_2$:	82050
"	— :	—			$Fe_2 Cl_6$:	196170
Zn	Zn O :	85430.	Zn S :	41989.	$Zn Cl_2$:	97210
Pb	Pb O :	50300.	Pb S :	19044.	$Pb Cl_2$:	82770
Ag	$Ag_2 O$:	5900.			Ag Cl :	29380
Hg	$Hg_2 O$:	42200.			Hg Cl :	82550
"	Hg O :	30660.			$Hg Cl_2$:	63160
An	— :	—			$An Cl_3$:	22820

In a Daniell's cell zinc is dissolved and copper precipitated; thus we have:

$$106090 - 55960 = 50130 \text{ h. u.}$$

All this heat cannot be converted into electricity; according to Kiliani only 83 per cent in the case of $Zn SO_4$ and 68 per cent in the case of $Cu SO_4$. The 50,130 h. u. are a measure of the electromotive power of the cell, and by dividing the numbers of heat units developed according to Thompson in the formation of the compounds enumerated above we find how many volts are required to decompose any of said compounds.

One Daniell cell is = 1.12 volt, the "volt" being the unit of electromotive power.

$$1 \text{ ampère} = \frac{1 \text{ volt}}{1 \text{ ohm}} =$$

= strength of current; the ampère — current precipitates in one minute 19.85 mg Cu and 67.57 mg Ag.

The Ohm is the unit of resistance, not quite accurately determined. The ohm is about = 95 per cent of a Siemens unit, or the resistance of a column of mercury 1 m. long and 1 sq. m. m. thick. The resistance is = $\frac{\text{length of conductor}}{\text{sectional area of conductor}}$ × the sp. resistance of the substance in question. The sp. res. of various elements are as follows, according to Matthiessen:

Cu	1.00		Pt	7.35
Ag	.77		Pb	9.96
Au	1.38		SC	18.07
Al	2.29		Hg	47.48
Zn	2.82		Bi	64.52
Fe	5.36		Graphite	1106.00
Sn	6.76		Gas-coke	2037.00

For actual metallurgical purposes galvanic elements are not sufficiently strong, and dynamo-electric machines have to be used in such cases.

1 weight of coal gives about 7500 h. u., but of these only about 360 can be converted into electricity in the dynamo-electric machine.

The working effect of a current per second is expressed by the product of the electromotive force in volts and the strength of current in ampères. One horse-power is = 75 meter-kilog. per second, and acceleration by gravitation is 9.81 m. If V = el. mot. power and A = strength of current, we have

$$\frac{A \times V}{9.81} = \text{electrical effect of a current in meter-kilog.,}$$

and

$$\frac{A \times V}{9.81 \times 75} = \text{the effect in horse-power.}$$

(For more complete information on electrical matters, the reader is referred to "Electricity and Electrical Engineering," by Fiske (Nostrand, New York).

The electrical units are briefly as follows:

$$1 \text{ dyne} = \frac{1}{981} \text{ of a gram.}$$
$$1 \text{ erg} = 1 \text{ dyne} - \text{centimetre.}$$

1 *electrostatic unit* of electricity is such a quantity as acts with one dyne on a similar quantity at a distance of one centimetre.

Current of unit strength is such a current as will act with the force of one dyne upon a magnet pole placed at the centre of a circle indicated by the conductor 1 centimetre long bent into an arc with 1 centimetre radius.

The unit quantity of current electricity is the quantity carried in one second by a current of unit strength (C. G. S. unit, centimetre, gram, second), *electromagnetic unit of electricity.*

The electromagnetic unit is much larger than the electrostatic unit, but as in both cases it should require 1 erg of work to give to each a unit of "potential," or electromotive force, it follows that the electromagnetic unit of electromotive force is much smaller than the electrostatic.

The following are the practical electromagnetic units, which are derived from the above C. G. S. units.

Electromotive force unit (difference of potential, potential) = *Volt* = 10^8 (C. G. S.) unit.
Resistance unit = *ohm* = 10^9 (C. G. S.) unit.
Strength of current unit = *ampère* = 10^{-1} (C. G. S.) unit.
Quantity of current unit = *coulomb* = 10^{-1} (C. G. S.) unit.
Unit of electrical work = volt × coulomb = *joule*.
Unit of electrical power = volt × ampère = *watt*.
1 *joule* = .1020 kilogram metres.
1 *watt* = $\frac{1}{746}$ horse-power = $\frac{1}{9.81}$ = .102 kilogram metres per second.

APPENDICES.

APPENDICES.

A.

HEAT CALCULATIONS.

(Heat unit = heat required to raise the temperature of 1 kilo. water 1° C.)

TABLE I. — *Gases at 0° and 760 Millimetres, according to Bunsen.*

	Sp. gr. referred to hydrogen.	Sp. gr. ref. to air.	Weight of 1,000 liters = 1 cubic metre kilo.	SPECIFIC HEAT.		HEAT UNITS DEVELOPED BY	
				Per 1 kilo.	Per 1 c. m.	Per 1 kilo.	Per 1 c. m.
Air,	$\begin{Bmatrix}O\\N\end{Bmatrix} 14.435$	1.000	1.2936	.2370	.3066	—	—
Oxygen,	$O = 16.0$	1.1056	1.4303	.2182	.3120	—	—
Nitrogen,	$N = 14.0$	.9713	1.2566	.2440	.3066	—	—
Hydrogen,	$H = 1.0$	.0692	.0896	3.4046	.3051	$\begin{Bmatrix}34462\\29633*\end{Bmatrix}$	$\begin{matrix}3088\\2655*\end{matrix}$
Carb. acid,	$\frac{CO_2}{2} = 22.0$	1.5202	1.9666	.2164	.4256	—	—
Carb. oxide,	$\frac{CO}{2} = 14.0$	.9674	1.2515	.2479	.3103	2403	3007†
Carbon gas,	$C = 12.0$	.8292	1.0727	—	—	—	—
Marsh gas,	$\frac{CH_4}{2} = 8.0$	.5531	.7155	.5929	.4242	11856*	8482*
Ethylene,	$\frac{C_2H_4}{2} = 14.0$	.9678	1.2520	.3694	.4625	11162*	13982*
Sulph. acid,	$\frac{SO_2}{2} = 32.0$	2.2139	2.8640	.1553	—	—	—
Sulphur. hy.,	$\frac{H_2S}{2} = 17.0$	1.1776	1.5234	.2423	—	2457	3743
Hydro. acid,	$\frac{HCl}{2} = 18.2$	1.2597	1.6296	.1845	.387	—	—
Steam,	$\frac{H_2O}{2} = 9.0$	.6221	.8048	.475	.3823	—	—

* The H burnt to H₂O, which passes off at 100°.
† 1 kilo. of C burnt to CO₂ gives 8080 heat unit.
" " " "   " "CO" 2473 " "
.5363 " " " "   " "CO₂" 4334 " "
" " " "   " "CO" 1327 " "
1 " " sulphur " "SO₂" 2221 " "

115

TABLE II. — *Number of Heat Units developed by Different Volumes of the Combustible Gases.* (*Vide* TABLE I.)

Volumes of gas, c. m.	Hydrogen, Heat Units.	Carbonic Oxide, Heat Units.	Marsh Gas, Heat Units.	Ethylene, Heat Units.
1	2655	3007	8482	13982
2	5310	6014	16964	27964
3	7965	9021	25446	41946
4	10620	12028	33928	55928
5	13275	15035	42410	69910
6	15930	18042	50892	83892
7	18585	21049	59374	97874
8	21240	24056	67856	111856
9	23895	27063	76338	125838

EXAMPLE:

```
Analysis:              Calorific effect:
CO₂ =  4.0 vol.        1 c. m. C₂H₄ gives            13982 heat units.
O   =  1.0  "          20 "  "  CO   "    60140 heat units
C₂H₄=  1.0  "          2  "  "  CO   "     6014   "    "
CO  = 22.0  "                                       ——————
H   =  5.0  "          5  "  "  H    "              66154  "    "
CH₄ =  1.0  "          1  "  "  CH₄  "              13275  "    "
N   = 66.0  "                                        8482  "    "
       ————               Total,                    ——————
      100.0                                         101893  "    "   per 100 c. m., or 1018.93
                                                           heat units per 1 c. m.
```

APPENDICES. 117

TABLE III. — *Number of Heat Units carried off by the Products of Combustion; the Gases being burnt with the Minimum Amount of Air, containing by Volume 1 Oxygen to 4 Nitrogen. The Flame Temperature, or Theoretical Temperature, for a Certain Gas is obtained by dividing the Calorific Power deducted from* TABLE II. *by the Number obtained from* TABLE III. *for the same Gas.*

Volumes of Gas in Cubic Metres	Heat Units carried off by the CO_2 already present in the Gas: Vol. CO_2 × .425.	Heat Units carried off by the Carbonic Acid and the Nitrogen resulting from the Combustion of the CO: Vol. CO × .425 + 2 vol. CO × .307 = 1.039.	Heat Units carried off by the Steam and Nitrogen from the Combustion of the Hydrogen: Vol. H × .382 + 2 vol. H × .307 = .996.	Heat Units carried off by the Carbonic Acid, the Steam and the Nitrogen from the Marsh Gas: Vol. CH_4 × .425 + 2 vol. CH_4 × .382 + 8 vol. CH_4 × .307 = 3.645.	Heat Units carried off by the Carbonic Acid, the Steam and the Nitrogen from the C_2H_4: 2 vol. C_2H_4 × .425 + 2 vol. C_2H_4 × .382 + 12 vol. C_2H_4 × .307 = 5.298.	Heat Units carried off by the Nitrogen already present in the Gas. Vol. N × .307.
1	.425	1.039	.996	3.645	5.298	.307
2	.850	2.078	1.992	7.290	10.596	.614
3	1.275	3.117	2.988	10.935	15.894	.921
4	1.700	4.156	3.984	14.580	21.192	1.228
5	2.125	5.195	4.980	18.225	26.490	1.535
6	2.550	6.234	5.976	21.870	31.788	1.842
7	2.975	7.273	6.972	25.515	37.086	2.149
8	3.400	8.312	7.968	29.160	42.384	2.456
9	3.825	9.351	8.964	32.805	47.682	2.763

The figures given in this table give the heat units carried off for each degree of temperature.

EXAMPLE. [*Compare Analysis and Calculation, p.* 88.]

```
4 vol. of CO₂ carry off heat units  . . . . . . .         1.700
20 vol of CO cause the carrying off of . . . . .  20.78
 2  "   "  CO    "      "       "   "  . . . . .   2.07
                                                          22.85
 5  "   "  H     "      "       "   "  . . . . . .        4.98
 1  "   "  CH₄   "      "       "   "  . . . . . .        3.64
 1  "   "  C₂H₄  "      "       "   "  . . . . . .        5.29
66  "   "  N — 4 (to correct for 1 vol. ox. present in the gas) 62 vol.  19.03
        Total heat units carried off per 100 c. m. of gas .  . .  57.490
```

Flame temperature = $\dfrac{101893}{57.49}$ = 1772° C.

TABLE IV.—*Number of Heat Units carried off by the Products of Combustion for each degree of temperature, the Gases being burnt with pure Oxygen; the theoretical temperature obtained as before.*

Volumes of Gas in Cubic Metres.	Heat Units carried off by the CO_2 already in the Gas: Vol. CO_2 × .425.	Heat Units carried off by the CO_2 resulting from the CO: Vol. CO × .425.	Heat Units carried off by the steam from the Hydrogen: Vol. H × .382.	Heat Units carried off by the CO_2 and the Steam from the Marsh Gas: Vol. CH_4 × (.425 + .307 × 2).	Heat Units carried off by the CO_2 and the Steam from the C_2H_4: 2 vol. C_2H_4 × .425 + 2 vol C_2H_4 × .382.	Heat Units carried off by the Nitrogen already present in the Gas: Vol. N × .307.
1	.425	.425	.382	1.189	1.414	.307
2	.850	.850	.764	2.378	2.828	.614
3	1.275	1.275	1.146	3.567	4.242	.921
4	1.700	1.700	1.528	4.756	5.656	1.228
5	2.125	2.125	1.910	5.945	7.070	1.535
6	2.550	2.550	2.292	7.134	8.484	1.842
7	2.975	2.975	2.674	8.323	9.898	2.149
8	3.400	3.400	3.056	9.512	11.312	2.456
9	3.825	3.825	3.438	10.701	12.726	2.763

Per kilog. oxygen { combining with or separating from } Fe to FeO { is developed or absorbed } 4732 heat units.
"   "   " {   "   "   } " " Fe_3O_4 "   "   4326 "   "
"   "   " {   "   "   } " " Fe_2O_3 "   "   4190 "   "
"   "   " {   "   "   } Mn " MnO   "   "   6875 "   "
"   "   " {   "   "   } " " MnO_2 "   "   4110 "   "
Iron   burnt to or reduced from Fe_2O_3 develops or absorbs 1796 "   "
"   "   "   "   " Fe_3O_4 "   "   "   1648 "   "
Manganese   "   "   "   " MnO_2 "   "   "   2410 "   "

EXAMPLE. [*Compare Analysis and Calculation, p. 88.*]

4 vol. of CO_2 carry off heat units 1.700
20 " " CO cause the carrying off of 8.5
2 " " CO "   "   "   " " 8
                                              ——
                                              9.300
5 " " H   "   "   "   " " 1.910
1 " " CH_4 " "   "   " " 1.189
1 " " C_2H_4 " "   "   " " 1.414
66 " " N—4 (to correct for 1 vol. ox. present in the gas) . . 19.030

Total heat units carried off for each degree of temperature . 34.543

$$\text{Flame temperature} = \frac{101893}{34.543} = 2950° \text{ C.}$$

APPENDICES.

TABLE V.—*Number of Heat Units developed by Different Weights of the Combustible Gases.* (*Vide* TABLE I.)

Weights of gas, kilog.	Hydrogen, Heat Units.	Carbonic Oxide, Heat Units.	Marsh Gas, Heat Units.	Ethylene, Heat Units.
1	29633	2403	11856	11162
2	59266	4806	23712	22324
3	88899	7209	35568	33486
4	118532	9612	47424	44648
5	148165	12015	59280	55810
6	177798	14418	71136	66972
7	207431	16821	82992	78134
8	237064	19224	94848	89296
9	266697	21627	106704	100458

EXAMPLE. [*Analysis, page 88.*]

CO_2 = vol. 4.0 ⎫
O = " 1.0
C_2H_4 = " 1.0   ⎬ To convert to % by weight, multiply the respective volumes with their respective specific gravity referred to hydrogen, *vide* Table I.
CO = " 22.0
H = " 5.0
CH_4 = " 1.0
N = " 66.0 ⎭
───
100.0

{ 4 × 22 = 88 = 6.4% weight of CO_2.
1 × 16 = 16 = 1.1% " " O.
1 × 14 = 14 = 1.0% " " C_2H_4.
22 × 14 = 308 = 22.6% " " CO.
5 × 1 = 5 = .3% " " H.
1 × 8 = 8 = .5% " " CH_4.
66 × 14 = 924 = 68.1% " " N.
─────
1363   100.0% }

Thus we find that 100 kilog. gas develop 78845 heat units. By a similar calculation to that on page 89, we find the flame temperature (correcting the nitrogen for the oxygen present by subtracting 1.1 × 3.3), which should agree with the temperature found the other way.

120   APPENDICES.

TABLE VI.—*Number of Heat Units carried off by the Products of Combustion for each degree of temperature; the Gases being burnt with the Minimum Amount of Air, containing 23.2 % of Oxygen and 76.8% of Nitrogen. Theoretical temperature obtained as before.*

CARBONIC ACID.	CARBONIC OXIDE.	HYDROGEN.	MARSH GAS.	ETHYLENE.	NITROGEN.
Weight $CO_2 \times .216$	$1.57 \times$ weight CO $\times .216 + 3.3 \times 1.57$ $\times$ weight CO $\times .244$	$9 \times$ weight $H \times .475$ $+ 3.3 \times 8 \times$ weight $H \times .244$	$2.25 \times$ weight CH_4 $\times .216 + 2.25 \times$ weight $CH_4 \times .475$ $+ 3.3 \times 4 \times$ weight $CH_4 \times .244$	$3.14 \times$ weight of $C_2H_4 \times .216 + 1.3 \times$ weight C_2H_4 $.475 + 3.3 \times C_2H_4$ $\times .244$	Weight $N \times .244$
.216	.797	10.446	3.882	4.027	.244
.432	1.594	20.892	7.764	8.054	.488
.648	2.391	31.338	11.646	12.081	.732
.864	3.188	41.784	15.528	16.108	.976
1.080	3.985	52.230	19.410	20.135	1.220
1.296	4.782	62.676	23.292	24.162	1.464
1.512	5.579	73.122	27.174	28.189	1.708
1.728	6.376	83.568	31.066	32.216	1.952
1.944	7.173	94.014	34.948	36.243	2.196

B.

CALCULATION OF BLAST-FURNACE BURDEN FROM THE ANALYSES OF ORES, FLUXES, AND FUEL-ASH, BY MEANS OF MRAZEK'S TABLE.

THE fluxing table, given on p. 92, is very simple and easily understood from a direct study of the same. The table shows at a glance the amounts required of each substance to give a slag of a certain composition. The equivalents in the last columns give the amounts of ore, etc., required to give one part of oxygen in excess of the respective proportions between the oxygen in the bases and the oxygen in the silica. Thus when the desired basicity or acidity has been obtained it is easy to find from the other columns the proportion between the alumina and the other bases, the proportion between lime and magnesia, the proportion between iron and slag, etc. If manganese is present, it is supposed that about one-half of the same goes into the slag, the other half being added to the iron. There is no need, in this calculation, of taking notice of other elements that are reduced and added to the iron in the blast-furnace.

Alumina makes a refractory slag, and promotes the formation of gray pig-iron; so does magnesia. *Much* slag also promotes the formation of gray pig-iron. In general, the proportion of alumina to the other bases ought not to exceed one to three. The relation of magnesia to lime should not be much above one to two. The relation of slag to iron varies between six tenths to one, and two to one.

When calculating the blast-furnace burden from careful chemical analyses, it should be borne in mind that the physical properties of the materials have much influence on the pig-iron resulting. Thus, for instance, very quartzy ores require. more heat to cause the silica to enter into chemical combination than do those ores that contain silica in chemical combination beforehand. Quartzy ores, therefore, promote the formation of gray pig-iron.

To introduce the ash of the fuel into our calculation, we first calculate the burden from the composition of the ores and the fluxes. Taking the consumption of coke at thirty-five parts per one hundred parts of ores and fluxes, we calculate how much coke is thus required for the quantities of ore and fluxes found from the tables of equivalents. Then we find (p. 74) how much ash this quantity of coke gives, and further, what part or multiple of a coke-ash equivalent this amount of ash forms. We have then only to introduce a corresponding part or multiple of some ore or flux equivalent to flux the coke-ash in concordance with the other burden.

APPENDICES.

FLUXING TABLE AT THE _____ BLAST-FURNACES.

	SLAG-FORMING INGREDIENTS.						IRON.	OXYGEN IN THE SLAG-FORMING INGREDIENTS.					EXCESS OF OXYGEN ABOVE						EQUIVALENTS, OR AMOUNTS OF ORE, ETC., REQUIRED TO GIVE 1 PART OF OXYGEN IN EXCESS OF:						
													Monosilicate 1 : 1.		Bisilicate 1 : 2.		1 : n Silicate.		Monosilicate 1 : 1.		Bisilicate 1 : 2.		1 : n Silicate.	
ORES, ETC.	MnO.	CaO.	MgO.	Al$_2$O$_3$.	SiO$_2$.	Total.		MnO.	CaO.	MgO.	Al$_2$O$_3$.	Oxygen in Bases "B."	Oxygen in SiO$_2$ "S."	In the Bases.	In the SiO$_2$.	In the Bases.	In the SiO$_2$.	In the Bases. β	In the SiO$_2$. γ	In the Bases.	In the SiO$_2$.	In the Bases.	In the SiO$_2$.	In the Bases.	In the SiO$_2$.
Iron ore .	.007	–	–	.065	.156	.228	.428	.0016	–	–	.0305	.0121	.0832	–	.0511	–	.0190	$\beta = B - \tfrac{1}{2}S$	$\gamma = S - nB$	–	19.59	–	52.62	1	γ
Iron ore .	.023	.159	.085	–	.011	.278	.226	.0051	.0454	.0340	–	.0845	.0059	.0786	–	.0815	–			12.72	–	12.27	–	1	–
Iron ore .	.003	.012	–	.010	.095	.120	.403	.0007	.0034	–	.0047	.0088	.0506	.1368	.0418	.1424	.0330			–	23.92	–	30.30	β	–
Limestone .	–	.518	–	–	.021	.539	–	–	.1480	–	–	.1480	.0112	–	.1388	–	–			7.31	–	7.02	–	1	γ
Coke-ash	–	.200	–	.150	.500	.850	.100	–	.0571	–	.0795	.1276	.2664	–	–	–	.0112			–	7.20	–	89.28		

1 part of CaO contains .286 part of oxygen. 1 part of Al$_2$O$_3$ contains .466 part of oxygen. 1 part of FeO contains .222 part of oxygen.
"   MgO   "   .400   "   "   "   SiO$_2$ "   .533   "   "   "   K$_2$O   "   .170   "   "
"   MnO   "   .225   "   "   "   Fe$_2$O$_3$ "   .300   "   "   "   Na$_2$O   "   .258   "   "

C.

TABLE FOR FACILITATING THE RAPID CALCULATION OF RESULTS OF ANALYSIS.

Weight of Precipitate obtained, Grams.	Five Grams of Sample taken [SiO_2] Si per cent.	Five Grams of Sample taken [CO_2] C per cent.	Five Grams of Sample taken [Mn_3O_4] Mn per cent.	Five Grams of Sample taken [$Mg_2P_2O_7$] P per cent.	Ten Grams of Sample taken [$BaSO_4$] S per cent.
.01	.093	.0545	.144	.056	.0137
.02	.186	.1090	.288	.112	.0274
.03	.279	.1635	.432	.168	.0411
.04	.372	.2180	.576	.224	.0548
.05	.465	.2725	.720	.280	.0685
.06	.558	.3270	.864	.336	.0822
.07	.651	.3815	1.008	.392	.0959
.08	.744	.4360	1.152	.448	.1096
.09	.837	.4905	1.296	.504	.1233

For example: in carbon determination, if the increase of the potash bulbs is found to be $= .1375$ grams, we have:

.5450
.1635
.0381
.0027

.7493 per cent of carbon

D.

ETCHING TEST FOR IRON AND STEEL.

Iron and steel surfaces are etched by means of a mixture of one part of strong HNO_3 with three parts of strong HCl, or two parts of strong HNO_3 with one part of concentrated H_2SO_4.

The acids attack the softer parts, and parts rich in slag, more vigorously than the metallic parts. The attacked parts appear soft and excavated.

The etched surface may be preserved by dipping it into lime-water after treating with acid, washing with water, and then applying a thin coating of wax.

E.

TABLE OF ELEMENTS, WITH SYMBOLS AND COMBINING WEIGHTS.

Name.	Symbol.	Com. Wt.	Name.	Symbol.	Com. Wt.
Aluminium	Al	27.4	Manganese	Mn	55
Antimony	Sb	122	Mercury	Hg	200
Arsenic	As	75	Molybdenum	Mo	96
Barium	Ba	137	Nickel	Ni	58.7
Bismuth	Bi	210	Niobium	Nb	94
Boron	B	11	Nitrogen	N	14
Bromine	Br	80	Osmium	Os	199.2
			Oxygen	O	16
Cadmium	Cd	112	Palladium	Pd	106.6
Cæsium	Cs	133	Phosphorus	P	31
Calcium	Ca	40	Platinum	Pt	197.5
Carbon	C	12	Potassium	K	39.1
Chlorine	Cl	35.5	Rhodium	Rh	104.4
Cerium	Ce	92	Rubidium	Rb	85.4
Chromium	Cr	52.5	Ruthenium	Ru	104.4
Cobalt	Co	58.7	Selenium	Se	79.5
Copper	Cu	63.5	Silver	Ag	108
Didymium	D	96	Silicon	Si	28
Erbium	E	112	Sodium	Na	23
			Strontium	Sr	87.5
Fluorine	F	19	Sulphur	S	32
Glucinum	Gl	9.5	Tantalum	Ta	172
Gold	Au	197	Tellurium	Te	129
Hydrogen	H	1	Thallium	Tl	204
			Thorium	Th	115.7
Indium	In	74	Tin	Sn	118
Iodine	I	127	Titanium	Ti	50
Iridium	Ir	198	Tungsten	W	184
Iron	Fe	56	Uranium	U	120
Lanthanum	La	92	Vanadium	V	137
Lead	Pb	207	Yttrium	Y	62
Lithium	Li	7	Zinc	Zn	65.2
Magnesium	Mg	24	Zerconium	Zr	89.6

APPENDICES.

F.

FRENCH WEIGHTS AND MEASURES,
AS MOST FREQUENTLY REQUIRED FOR ENGLISH COMPARISONS.

WEIGHTS.

1 milligramme = .015438 English Troy grains.
1 gramme   = 15.438   "   "   "
           or .002205 of a lb. Avoirdupois.
1 kilogramme = 2.2048 lbs. Avoirdupois.

WEIGHTS AS POPULARLY ESTIMATED.

	lbs.	ozs.	drs. Avoirdupois.
1 gramme ... =	0	0	0¼
1 decagramme .. =	0	0	5¾
1 hectogramme . =	0	3	8¼
1 kilogramme .. =	2	3	4¼

LINEAL MEASURES.

1 millimetre ... = .039371   English inches.
25¼   "   ... = 1   "   inch.
1 metre   ... = 39.371 or 39⅜   "   inches.
       or   = 3.2809 feet = 3⅜ inches.

MEASURES OF SURFACE.

1   centiare   = 1.196 English square yards, or 10.764 English sq. ft.
9.3 centiares = 100   "   "   feet.
1   are   = 119.6   "   "   yards.
40.47 ares   = 1   "   statute acre.
1   hectare   = 2.47   "   "   "

MEASURES OF SOLIDITY.

1 millistere . . . . = .035317 English cubic feet.
1 stere . . . . = 35.317   "   "   "

MEASURES OF CAPACITY.

1 litre . . . . = 61.028 English cubic inches.

POPULAR MEASURES OF CAPACITY.

	English galls.	qts.	pts. imp.
1 litre . . . . =	0	0	1¾ "
1 decalitre . . . . =	2	0	1½ "
1 hectolitre . . . . =	22	0	0 "
1 kilolitre . . . . =	220	0	0 "

CONVERSION OF EQUIVALENT MEASURES.

ENGLISH TO FRENCH.			FRENCH TO ENGLISH.		
Inches × .0254	=	metres	× 39.371	=	inches.
Feet × .30477	=	"	× 3.2809	=	feet.
Yards × .91438	=	"	× 1.09364	=	yards.
Miles × 1.6093	=	kilometres	× .62138	=	miles.
Acres × .40467	=	hectares	× 2.4712	=	acres.
Imp. galls. × 4.54339	=	litres	× .2201	=	gallons.
Cubic inches × .01639	=	"	× 61.028	=	cubic ins.
Bushels × .36347	=	hectolitres	× 2.75125	=	bushels.
Quarters × 2.9077	=	"	× .3439	=	quarters.
Troy grs. × .06479	=	grammes	× 15.434	=	Troy grs.
Troy lbs. × .3732	=	kilogrammes	× 2.6795	=	Troy lbs.
Avoir. lbs. × .4535	=	"	× 2.2048	=	Avoir. lbs.

G.

"BODY" IN STEEL.

It has been pointed out in the first chapter of this book how difficult it is even with the use of the comparatively accurate analyses of the present time to find the true connection between chemical composition and physical properties of steel. The influence of minute quantities of impurities has been demonstrated in many metals besides iron, by various scientists. It is quite possible that the presence of minute amounts of such elements that are not included in what is generally termed a "complete analysis" of steel may cause some of the wide differences in physical properties which are sometimes observed in otherwise similar steels.

Be this as it may, there is something in steel for which we cannot account, and "body" it is called in Sheffield and in Sweden. The cast-steel made from the best Swedish steel-iron has more "body" than other cast-steels; it will stand a larger number of heatings, etc.

But it is not only the crucible steel that has need of the word "body." Take for instance the soft Bessemer steel and Martin steel, at present made in Sweden, from the same materials as are used for the famous steel-irons. Horseshoe nails, rivets, ship plates, etc., from this soft steel show an endurance in service, a "body," that can rarely be attained by steels made from other materials and "doctored" in an open hearth furnace so as to have nearly the same chemical composition. Some manufacturers using only wrought-iron claim that the softest cast-metal cannot stand the constant vibrations sufficiently, whilst others have abandoned the wrought-iron in favor of the soft metal made from the "steel-iron" ores.

The same "body" is noticed in the harder varieties of steel

made from the old "steel-iron" ores. No satisfactory explanation regarding this has yet been reached; but it is hoped that experiments such as those conducted by Mr. Brinell (*vide* above) will throw more light on the subject.

The following analyses show the composition of a soft Bessemer steel made from the Swedish steel-iron ores and a steel made from cheaper ores; the steels are both used for nails, etc.

1. Swedish Ore.	2. Cheaper Ore.
P = .027	P = .065
Mn = .14	Mn = .51
Si = .01	Si = .01
S = .01	S = .04
C = .10–.15	C = .11

The former steel shows a decidedly better chemical composition than the latter, but even when the compositions happen to be nearly alike, the "body" will assert itself.

H.

MELTING POINTS, SPECIFIC HEAT, LATENT FUSING HEAT AND ELECTRICAL CONDUCTIVITY OF METALS.

Metals.	Melting point. Deg. C.	Specific heat.	Latent fusing heat.	Conductivity Hg = 1.
Al	700	.21224		31.7
Sb	450	.04989		2.0
As		.0758		2.6
Pb	334	.0402 fluid	5.858	4.8
Cd	320	.0548	13.660	13.9
Soft iron	1600	.10808		8.3
Pig-iron, white	1050–1100		23 000	
" gray	1200–1300		33.000	
Steel	1300–1400	.1175		
Au	1075	.0316		43.8
Co	1800	.10674		9.6
Cu	1200	.09483	121.2	52.2
Mn	1900	.1217		
Ni	1600	.10916		7.3
Osmium	2500	.03113		
Palladium	1500	.0592	36.30	6.9
Platinum	1775	.0377	27.18	8.2
Hg	−39	.0335	2.82	1.0
Rhodium	2000	.05803		
Ruthenium	1800	.0611		
Iridium	2200	.0323		
Ag	954	.0559	21.07	57.2
Bi	260	.0363 fluid	12.64	.8
Zn	412	.0932	28.13	16.1
Sn	228	.0637 fluid	13.31	8.2
Cr	Higher than Pt.	.0997		
Wo	" " "	.0350		

I.

The accompanying list of Apparatus and Reagents is intended to give some idea as to what is wanted in setting up a laboratory for regular steel works analyses; the idea of putting down the prices was only a second thought. As it is impossible, on account of the fluctuating prices of articles like these, to be exactly right, it may seem rather out of the way to give any price at all; but it is done so as to give some idea as to what such a laboratory could be stocked for, and as the highest price has been given for everything, especially the apparatus, it would perhaps be nearer the right amount if the total were to be discounted, say, 25 per cent, and the place could be fitted up for, say, $389.00; this does not include the platinum combustion apparatus given in this book, as it stands, not counting oxygen receiver, it would cost about $100.00. In round numbers it would cost $500.00; this would be a well-fitted-up laboratory ready to turn out any kind of work usually done in a steel works, and with the stock on hand should run from four to eight months, according to the amount of work required to be turned out in that time. Of course the most useful reagents, such as the acids and ammonium hydrate, might be required before that, and then other chemists might want other things which are not mentioned in this list, and which they would consider highly important to have.

REAGENTS.

Acid, acetic, 5 lbs. @ .13	$ .65
" chromic, 1 lb.	2.00
" hydrochloric, 30 lbs. @ .25	7.50
" hydrofluoric, ¼ lb.	1.25
" molybdic, 3 lbs. @ 5.00	15.00
" nitric, 49 lbs. @ .25	12.25

APPENDICES.

Acid, oxalic, ¼ lb.. $	.40
" sulphuric, 18 lbs. @ .25..	4.50
Alcohol, 1 pt..	.60
Ammonium chloride, 2 lbs. @ .50	1.00
" hydrate, 28 lbs. @ .15....................................	4.20
" oxalate, ¼ lb..	.35
" carbonate, ½ lb...	.35
Asbestos, 2 lbs. @ 1.00...	2.00
Barium carbonate, ½ lb..	1.70
" chloride, 1 lb...	.30
Bromine, 1 lb...	.80
Calcium chloride (dry), 2 lbs. @ .75................................	1.50
Copper and ammonic chloride, 20 lbs	18.00
" sulphate (common), 5 lbs. @ .10............................	.50
Iron sesquichloride, 1 lb	1.00
" protosulphate, 5 lbs. @ .10	.50
" " and ammon., 1 lb..	.20
" sulphide, 2 lbs. @ .20..	.40
Iodine, 1 oz...	.30
Magnesium chloride, 1 lb...	.50
Mercury, 1 lb...	.65
Microcosmic salt, ½ lb..	.70
Paper, white wrapping, 5 qrs. @ .30................................	1.50
" filtering schleicher and schulls, 595, 5 qrs. @ .60..............	3.00
" " " " " 589 cut, C. P. 11 C. M. 3 pck. @ 1.10	3.30
" " " " " " " " 9 " 2 " @ .90	1.80
" " " " " " " " 7 " 4 " @ .70	2.80
Potassium carbonate, ½ lb	.50
" bi-chromate, 2 lbs. @ .50...................................	1.00
" chlorate, 5 lbs. @ .50.....................................	2.50
" cyanide-ferro, ¼ lb..	.25
" " ferri, ¼ lb..	.50
" hydrate (by alcohol), 2 lbs. @ 1.50.........................	3.00
" permanganate, 1 lb	.70
Pumice stone, 1 lb..	.10
Silver nitrate, ½ oz..	.50
Sodium acetate, 1 lb..	.75
" carbonate (dry C. P.), 2 lbs. @ .60..........................	1.20
" thiosulphate, 2 lbs. @ .60	1.20

134   APPENDICES.

Zinc, metallic C. P. ½ lb. .90, 2 lbs. common, @ .25, .50	$ 1.40
Wax, bees', ½ lb.	.30
" sealing, ¼ lb.	.20

$104.60

APPARATUS.

1 Balance, analytical, to weigh up to 100 grms. and sensible to $\frac{1}{10}$ milligramme	$ 95.00
1 Balance, Robervahl No. 3	7.50
1 set weights, 50 grms. and down to 1 milligramme, with 3 riders	16.00
1 " " 1 kilo. to 1 gramme	8.50
Beakers, Griffin's wide form, lipped, 1-6; 12 nests, @ $1.85	22.20
" " " " " 1-3; 12 " @ .60	7.20
" conical Bohemian glass, ½ litre capacity, 12 @ .35	4.20
Bellows, Fletcher's new pattern No. 9, a foot blower	6.00
Blast Lamp, Bunsen No. 5	3.50
Burners, Fletcher's solid flame No. 47, large size, 3 @ 2.00	6.00
" Finckner's improved form, 6 @ 2.00	12.00
Bottles, Reagent, 1 pint glass stoppered, 1 dozen	2.00
" " 1 quart " " 1 "	2.50
" common wide mouth, for drillings, 2 oz., 4 gross @ 4.20	16.80
" Woulff's 2 necked, pint, 4 @ .50	2.00
Brushes, camel's hair, small, 6	.15
" " " large, 1	.05
Bulbs, for sulphur determination, Troilius improved form, 6	4.50
Burettes, Mohr's, 50 c.c. grad. to $\frac{1}{10}$ c.c., 1	1.50
' " 100 c.c. " to $\frac{1}{10}$ c.c., 1	2.50
Carboys for distilled water, 2 @ 1.50	3.00
Condenser, block tin worm, in zinc tank, to be attached to steam pipe	8.00
Corks, wooden, to fit drilling bottles, 4 gross @ 2.00	8.00
" " assorted, small, 1 gross	3.50
" rubber " 1 lb.	3.00
Crucibles, porcelain, Royal Berlin, No. 00, 6 @ .18	1.08
" platinum, each 20 gr. 4 = 80 grammes @ .35	28.00
Cylinders, Erdmann, 4 @ .25	1.00
" glass, 1000 c.c. grad. to 10 c c., ground stopper, 1	3.00
" " 25 c.c. " " 1 c.c., 1	.25

APPENDICES. 135

Dishes, porcelain, pint, 4 @ .50	$ 2.00
" " 4 oz. 4 @ .25	1.00
Files, 4" △, 4 @ .12	.48
Flasks, 2 gallon, 1	2.00
" 2 qt., 6 @ .50	3.00
" 1 " 12 @ .35	4.20
" 1 pt., 6 @ .25	1.50
" 1 " ground stopper, labeled from A to L, 12	12.25
Funnels, 2½ inches, 2 dozen @ 2.25	4.50
" 1½ " 2 " @ 1.75	3.50
" 4 " 4 @ .18	.72
" 9 " 2 @ .35	.70
" for filtering with asbestos, 12 @ .25	3.00
Glass, measuring 16. oz., graduated	.62
" covers, Watch, 2½", 12	1.25
" " " 4", 12	2.00
" " " 5", 12	2.50
Hydrometers, one for above 1.000 and one for below 1.000 @ 1.00	2.00
" jars, 18" × 1¾"	.60
Labels, gummed, No. 221, 12 boxes @ .10	1.20
" Reagent, 2 books, @ .25	.50
Mortar, agate, 3½"	8.00
" Wedgewood, 3"	.40
" " 5½"	.80
" iron, 2 gallon	4.00
Tongs, crucible	1.50
Pipette, 100 c. c.	.70
" 50 c. c., 2 @ .50	1.00
" 10 c. c.	.25
Platinum triangles, 4 each 11 grammes — 44 grammes, @ .35	15.40
Plates, porcelain, for testing, 3 @ .65	1.95
Racks, test tube, 2 @ .75	1.50
Sieves, copper, with mesh, 50 to 1 in., and 80 to 1 in., 2 @ .50	1.00
Stands, burette, 1	1.25
" filter, 6 @ 1.00	6.00
" retort (iron), with rings, 6 @ 1.25	7.50
" tripod, 4 @ .50	2.00
Thermometers, 360° Cent., 2 @ 2.00	4.00
Rubber finger tips, 1 doz.	.50

APPENDICES.

Tubes, aspirator, Geissler's, 2 @ .75	$ 1.50
" funnel, stopcock, 2 @ 1.00	2.00
" test, 5" × ⅝", 5 doz., @ .30	1 50
" for carbon determination, 1 set	2.50
" glass, assorted, 3 lbs. @ .50	1 50
Rods, " " 3 lbs. @ .50	1.50
Rubber tubing, vulcanized, ¼", 24 ft., @ .15	3.60
" " pure, ⅛", 12 ft., @ .10	1.20
" " " ³⁄₁₆, 12 ft., @ .14	1.68
1 Desiccator	4.00

SUNDRIES.

1 broom, .35; 1 dust brush and pan, .50; scissors, .75; hammer, 1.25	2.85
1 small magnet, .25; clock, 2.00; camera for c. determination, 2.50	4.75
2 baths for c. determination, 2.00; matches, .25	2.25
Total for Apparatus	414.03
" " Reagents	104.60
	$518.63

J.

LIST OF THE PRINCIPAL IRON ORES.

Carbonates of iron (with carbonates of Mn, Ca and Mg):
Siderite, chalybite, spathic and sparry iron, steel ore.
Spherosiderite, often containing manganese.
Sideroplesite.
Siderodot.
The crystallized carbonates in layers.
Earthy, lithoid, argillaceous, and calcareous carbonates of iron.

Clay Ironstone (common name given to many different kinds of iron ores). (Fe_2O_3, Al_2O_3, HO).
Nodular carbonates.
Black band iron-stone of England.
Cleveland iron-stone of England.

Magnetic Ores (TiO gangue, etc.).
Magnetite.
Ochreous magnetite.
Magnetic sands.

Red Hematites (H_2O, gangue MnO, CuO, Al_2O_3, Si_3, P_2O_5, TiO_2):
Specular iron, oligist iron, iron glance.
Micaceous iron.
Martite.
Violet ore.
Red ochre.
Specular schist.
Clay iron-stone, argillaceous hematite.
Puddlers' ore from Cumberland, England.
Lenticular iron ore, oolitic fossil ore.

Brown hematites: Turgite.

Göthite, pyrrhosiderite, brown iron-stone or ore.
Limonite, brown ochre.
Xanthosiderite.

Franklinites: ($Fe_3 O_3$, Mn O, Ca O, Si O_3, Mg O, Zn O).

For the analysis of chrome-iron-ores and the determination of many rare elements by convenient methods, the reader is referred to Roscoe-Schoslemmer's Chemistry, Post and others. It is beyond the province of this book to enter upon such analyses as do not occur in the ordinary routine of an Iron and Steel Laboratory.

For examples of analyses of iron ores, *vide* the Records of the Geological Survey of Pennsylvania.

K.

Determination of organic matter in water for drinking purposes. By Prof. Eggertz.

Requisites:

= A beaker 10 c. m. high, 5 c. m. wide, with a 100 c. c. mark.

= A measuring test tube, 10 c. m. long and .9 c. m. inside diameter; graduated into 1 and ½ c. c.

= A glass rod 12 c. m. long.

= Pure concentrated sulphuric acid.

= Solution of .435 gram pure, dry, cryst. permanganate of potassium in 1000 c. c. of distilled water. Every c. c. of this solution contains .1 milligramme oxygen for the destruction of organic substances. The organic substances may be of many different kinds, requiring different amounts of oxygen, but it is supposed that the total quantity of these substances is = 20 times the amount of oxygen consumed.

100 c. c. of the water for testing are taken into the abovementioned beaker, 5 c. c. of $H_2 SO_4$ and 3 c. c. of $K Mn O_4$ are added. The beaker is then put into a pot with boiling water; on the bottom of the pot a piece of cloth is placed for the beaker to rest upon. The beaker is steadied by means of a triangle of brass or iron wire. The boiling water should be on the same level as the water in the beaker. After about 3 minutes the water in the beaker has a temperature of about 90°, and retains this as long as the water outside is kept at a gentle boil. If after 5 minutes at 90° C. a red color remains the water may be considered as good. If the color disappears another 3 c. c. $K Mn O_4$ are added. If now after 5 minutes at 90° C. the red color remains, the water *may be* used for drink-

ing purposes. By continuing in this manner the quality of any water may be shown.

The volume of the water in the beaker should be kept constant by adding a little distilled water now and then, as at 90° C. about 1 c. c. is evaporated every minute.

L.

TABLE GIVING THE TENSION OF AQUEOUS VAPOR IN M. M. MERCURY BETWEEN 0° AND 40° C., ACCORDING TO REGNAULT.

$\log. p = a + b\alpha^{t^o} + c\beta^{t^o}.$

$a = 4.7393707.$

$\log. (b\alpha^{t^o}) = + 0.6117408 — 0.003274463\ t^o,$ (b negative).

$\log. (c\beta^{t^o}) = — 1.8680093 + 0.006864937\ t^o,$ (c positive).

t_o	p.	$\log. p$.	t_o	p.	$\log. p$.
0	4.600	0.6627572	21	18.494	1.2670381
1	4.940	0.6936961	22	19.659	.2935630
2	5.302	.7244076	23	20.888	.3198954
3	5.687	.7548934	24	22.184	.3460375
4	6.098	.7851559	25	23.550	.3719908
5	6.534	.8151954	26	24.989	.3977556
6	6.999	.8450155	27	26.505	.4233355
7	7.492	.8746153	28	28.102	.4487305
8	8.017	.9040012	29	29.781	.4739419
9	8.574	.9331711	30	31.548	.4989725
10	9.165	.9621263	31	33.406	.5238222
11	9.792	.9908709	32	35.359	.5484945
12	10.457	1.0194058	33	37.410	.5729897
13	11.162	.0477316	34	39.565	.5973082
14	11.908	.0758510	35	41.827	.6214535
15	12.699	.1037660	36	44.200	.6454260
16	13.536	.1314767	37	46.690	.6692267
17	14.421	.1589861	38	49.301	.6928579
18	15.357	.1862950	39	52.038	.7163212
19	16.346	.2134062	40	54.906	.7396162
20	17.391	.2403204			

M.

USEFUL TABLES.

TABLE OF THE WEIGHT OF BAR IRON.

SQUARE.

Side of square in inches.	Length, one foot.		Side of square in inches.	Length, one foot.	
	Weight in pounds.	Cubic weight in pounds.		Weight in pounds.	Cubic weight in pounds.
1/4	.209	.0174	2 1/8	15.083	1.257
3/8	.470	.0391	2 1/4	16.909	1.409
1/2	.835	.0696	2 3/8	18.840	1.570
5/8	1.305	.1087	2 1/2	20.875	1.739
3/4	1.879	.1565	2 5/8	23.115	1.926
7/8	2.558	.2131	2 3/4	25.259	2.105
1 in.	3.340	.2783	2 7/8	27.608	2.301
1 1/8	4.228	.3523	3 in.	30.070	2.506
1 1/4	5.219	.4349	3 1/4	35.279	2.940
1 3/8	6.315	.5262	3 1/2	40.916	3.409
1 1/2	7.516	.6263	3 3/4	46.969	3.914
1 5/8	8.820	.7350	4 in.	53.440	4.455
1 3/4	10.229	.8524	4 1/2	67.637	5.636
1 7/8	11.743	.9786	5 in.	83.510	6.959
2 in.	13.360	1.113			

TABLE OF SPECIFIC GRAVITIES.

Metals.	Weight, water being 1000.	Number of cubic inches in a pound.	Weight of a cubic inch in pounds.
Platina	19500	1.417	.7053
Pure gold	19258	1.435	.6965
Mercury	13560	2.04	.4902
Lead	11352	2.435	.4105
Pure silver	10474	2.638	.3788
Bismuth	9823	2.814	.3552
Copper	8788	3.146	.3178
Brass	7824	3.533	.3036
Iron, cast	7264	3.806	.263
Iron, bar	7700	3.592	.279
Steel	7833	3.530	.2833
Tin	7291	3.790	.2636
Zinc	7190	3.845	.26

APPENDICES.

APPROXIMATE WEIGHTS FOR PRACTICAL PURPOSES.

	Specific gravities.	Weight of a cubic inch in pounds.	Weight of a cylindrical inch in pounds.
Iron, cast............	7207	.2608	.2048
Iron, bar.............	7700	.2785	.2187
Steel.................	7833	.2833	.2225
Copper...............	8878	.3211	.2522
Brass, cast...........	8396	.3037	.2385
Zinc, common........	7028	.2542	1996
Lead.................	11352	.4106	.3225
Tin, cast.............	7291	.2637	.2071

TABLE OF THE WEIGHT OF FLAT BAR, HOOP, PLATE, AND SHEET IRON.

Weight of a Lineal Foot of Flat Bar and Hoop Iron in pounds.

Thickness in inches.	Breadth in inches.									
	$3\frac{1}{2}$	3	$2\frac{3}{4}$	$2\frac{1}{2}$	$2\frac{1}{4}$	2	$1\frac{3}{4}$	$1\frac{1}{2}$	$1\frac{1}{4}$	1	$\frac{3}{4}$
$\frac{1}{8}$	1.47	1.26	1.15	1.05	.094	.084	.073	.063	.052	.042	.031
$\frac{3}{16}$	2.20	1.89	1.73	1.57	1.41	1.26	1.10	.094	.078	.063	.047
$\frac{1}{4}$	2.94	2.52	2.31	2.10	1.89	1.68	1.47	1.26	1.05	.084	.063
$\frac{3}{8}$	4.41	3.78	3.46	3.15	2.83	2.52	2.20	1.89	1.57	1.26	094
$\frac{1}{2}$	5.88	5.04	4.62	4.20	3.78	3.36	2 94	2.22	2.10	1.68	1.26
$\frac{5}{8}$	7.35	6.30	5.77	5.25	4.72	4.20	3.67	3.15	2.62	2.10	1.57
$\frac{3}{4}$	8.82	7 56	6.93	6.30	5.66	5.04	4.41	3.78	3.15	2.52	
$\frac{7}{8}$	10.29	8.82	8.08	7.35	6.61	5.88	5.14	4.41	3.67	2.94	
1 in.	11.76	10 08	9.24	8.40	7.56	6.72	5.87	5.04	4.20		

Weight of a Square Foot of Plate Iron in pounds.

Thickness in parts of an inch....	$\frac{1}{8}$	$\frac{3}{16}$	$\frac{1}{4}$	$\frac{5}{16}$	$\frac{3}{8}$	$\frac{7}{16}$	$\frac{1}{2}$	$\frac{9}{16}$	$\frac{5}{8}$	$\frac{11}{16}$	$\frac{3}{4}$
Weight it pounds...............	5	$7\frac{1}{2}$	10	$12\frac{1}{2}$	15	$17\frac{1}{2}$	20	$22\frac{1}{2}$	25	$27\frac{1}{2}$	30

Weight of Square Foot of Sheet Iron in pounds.

Number on wire gauge, and weight in pounds.	1	2	3	4	5	6	7	8	9	10	11
	12.5	12	11	10	9	8	7.5	7	6	5.68	5
Number on wire gauge, and weight in pounds.	12	13	14	15	16	17	18	19	20	21	22
	4.62	4.32	4	3.95	3	2.5	2.18	1.93	1 62	1.5	1.37

BOOKS OF REFERENCE.

LEDEBUR, *Eisenhüttenkunde.*
EGGERTZ, *Jernkontorets Annaler.*
       Transactions American Institute Mining Engineers.
       Jernkontorets Annaler.
       Journal British Iron and Steel Institute.
POST, *Chemische Untersuchungs Methoden.*
THORPE, *Quantitative Analysis.*
BAILEY, *Chemists' Pocketbook.*
BALLING, *Metallhüttenkunde.*
HOARE, *Iron and Steel.*
JÜPTNER VON JONSTORFF, *Eisenhüttenchemiker.*
BECKERT, *Eisenhüttenkunde.*
FRED. TAYLOR, *Ziemens Gas vs. Watergas.*

LIST OF CHEMISTS,

WHOSE NAMES ARE ASSOCIATED WITH METHODS GIVEN IN THIS BOOK.

Carbon by Combustion. . . . . McCreath, Ullgren.
" " *Color* . . . . Eggertz, Jenkins, Snelus.
Phosphorus . . . . Tamm, Snelus.
Manganese. . . . . Beilstein and Jawein, Pattinson, Ford.
Iron by Ki and Hg . . . . Morrell, Thorpe.
Arsenic . . . . Lundin.
Tungsten. . . . . Eggertz.
Gas Analysis. . . . . Bunte, Eggertz, and others.

www.ingramcontent.com/pod-product-compliance
Lightning Source LLC
Chambersburg PA
CBHW030337170426
43202CB00010B/1156